FRONTIERS OF SPACE EXPLORATION

Other Titles in the Greenwood Press Guides to Historic Events of the Twentieth Century
Randall M. Miller, Series Editor

The Persian Gulf Crisis
Steve A. Yetiv

World War I
Neil M. Heyman

The Civil Rights Movement
Peter B. Levy

The Holocaust
Jack R. Fischel

The Breakup of Yugoslavia and the War in Bosnia
Carole Rogel

Islamic Fundamentalism
Lawrence Davidson

FRONTIERS OF SPACE EXPLORATION

Roger D. Launius

Greenwood Press Guides to
Historic Events of the Twentieth Century
Randall M. Miller, Series Editor

Greenwood Press
Westport, Connecticut • London

Library of Congress Cataloging-in-Publication Data

Launius, Roger D.
Frontiers of space exploration / Roger D. Launius.
p. cm.—(Greenwood Press guides to historic events of the twentieth century, ISSN 1092–177X)
Includes bibliographical references and index.
ISBN 0–313–29968–4 (alk. paper)
1. Outer space—Exploration. 2. Astronautics—International cooperation. I. Title. II. Series.
QB500.262.L38 1998
629.45—dc21 97–34788

British Library Cataloguing in Publication Data is available.

Library of Congress Catalog Card Number: 97–34788
ISBN: 0–313–29968–4
ISSN: 1092–177X

First published in 1998

Greenwood Press, 88 Post Road West, Westport, CT 06881
An imprint of Greenwood Publishing Group, Inc.

Printed in the United States of America

The paper used in this book complies with the Permanent Paper Standard issued by the National Information Standards Organization (Z39.48–1984).

10 9 8 7 6 5 4 3

Front cover photo: *Apollo 14*, *Saturn V* space vehicle lifting off to the moon, January 1971. (NASA photo number 71-H-221)

Back cover photo: The Hubble Space Telescope, launched in 1990 and repaired in 1993. (NASA photo number 94-H-16)

For Lee D. Saegesser,
NASA chief archivist for more than thirty years

ADVISORY BOARD

Contents

Series Foreword

As the twenty-first century approaches, it is time to take stock of the political, social, economic, intellectual, and cultural forces and factors that have made the twentieth century the most dramatic period of change in history. To that end, the Greenwood Press Guides to Historic Events of the Twentieth Century presents interpretive histories of the most significant events of the century. Each book in the series combines narrative history and analysis with primary documents and biographical sketches, with an eye to providing both a reference guide to the principal persons, ideas, and experiences defining each historic event, and a reliable, readable overview of that event. Each book further provides analyses and discussions, grounded in both primary and secondary sources, of the causes and consequences, in thought and action, that give meaning to the historic event under review. By assuming a historical perspective, drawing on the latest and best writing on each subject, and offering fresh insights, each book promises to explain how and why a particular event defined the twentieth century. No consensus about the meaning of the twentieth century emerges from the series, but, collectively, the books identify the most salient concerns of the century. In so doing, the series reminds us of the many ways those historic events continue to affect our lives.

Each book follows a similar format designed to encourage readers to consult it both as a reference and a history in its own right. Each volume opens with a chronology of the historic event, followed by a narrative overview, which also serves to introduce and examine briefly the main themes and issues related to that event. The next set of chapters is composed of topical essays, each analyzing closely an issue or problem of interpretation introduced in the opening chapter. A concluding chapter suggesting the long-term implications and meanings of the historic event brings the strands of the preceding chapters together while placing the event in the larger historical context. Each book also includes a section of short biographies of the principal persons related to the event, followed by a section introducing and reprinting key historical documents illustrative of and pertinent to the event. A glossary of selected terms adds to the utility of each book. An annotated bibliography—of significant books, films, and CD-ROMs—and an index conclude each volume.

The editors made no attempt to impose any theoretical model or historical perspective on the individual authors. Rather, in developing the series, an advisory board of noted historians and informed high school history teachers and public and school librarians identified the topics needful of exploration and the scholars eminently qualified to examine those events with intelligence and sensitivity. The common commitment throughout the series is to provide accurate, informative, and readable books, free of jargon and up to date in evidence and analysis.

Each book stands as a complete historical analysis and reference guide to a particular historic event. Each book also has many uses, from understanding contemporary perspectives on critical historical issues, to providing biographical treatments of key figures related to each event, to offering excerpts and complete texts of essential documents about the event, to suggesting and describing books and media materials for further study and presentation of the event, and more. The combination of historical narrative and individual topical chapters addressing significant issues and problems encourages students and teachers to approach each historic event from multiple perspectives and with a critical eye. The arrangement and content of each book thus invite students and teachers, through classroom discussions and position papers, to debate the character and significance of great historic events and to discover for themselves how and why history matters.

The series emphasizes the main currents that have shaped the modern world. Much of that focus necessarily looks at the West, especially Europe and the United States. The political, commercial, and cultural expansion of the West wrought largely, though not wholly, the most fundamental changes of the century. Taken together, however, books in the series reveal the interactions between Western and non-Western peoples and society, and also the tensions between modern and traditional cultures. They also point to the ways in which non-Western peoples have adapted Western ideas and technology and, in turn, influenced Western life and thought. Several books examine such increasingly powerful global forces as the rise of Islamic fundamentalism, the emergence of modern Japan, the communist revolution in China, and the collapse of communism in eastern Europe and the former Soviet Union. American interests and experiences receive special attention in the series, not only in deference to the primary readership of the books but also in recognition that the United States emerged as the dominant political, economic, social, and cultural force during the twentieth century. By looking at the century through the lens of American events and experiences, it is possible to see why the age has come to be known as "The American Century."

Assessing the history of the twentieth century is a formidable prospect. It has been a period of remarkable transformation. The world broadened and narrowed at the same time. Frontiers shifted from the interiors of Africa and Latin America to the moon and beyond; communication spread from mass circulation newspapers and magazines to radio, television, and now the Internet; skyscrapers reached upward and suburbs stretched outward; energy switched from steam, to electric, to atomic power. Many changes did not lead to a complete abandonment of established patterns and practices so much as a synthesis of old and new, as, for example, the increased use of (even reliance on) the telephone in the age of the computer. The automobile and the truck, the airplane, and telecommunications closed distances, and people in unprecedented numbers migrated from rural to urban, industrial, and ever more ethnically diverse areas. Tractors and chemical fertilizers made it possible for fewer people to grow more, but the environmental and demographic costs of an exploding global population threatened to outstrip natural resources and human innovation. Disparities in wealth increased, with developed nations prospering and underdeveloped nations starving. Amid the crumbling of former European colonial empires, Western technology, goods, and culture increasingly enveloped the globe, seeping into, and undermining, non-Western cultures—a process that contributed to a surge of religious fundamentalism and ethno-nationalism in the Middle East, Asia, and Africa. As people became more alike, they also became more aware of their differences. Ethnic and religious rivalries grew in intensity everywhere as the century closed.

The political changes during the twentieth century have been no less profound than the social, economic, and cultural ones. Many of the books in the series focus on political events, broadly defined, but no books are confined to politics alone. Political ideas and events have social effects, just as they spring from a complex interplay of non-political forces in culture, society, and economy. Thus, for example, the modern civil rights and women's rights movements were at once social and political events in cause and consequence. Likewise, the Cold War created the geopolitical framework for dealing with competing ideologies and nations abroad and served as the touchstone for political and cultural identities at home. The books treating political events do so within their social, cultural, and economic contexts.

Several books in the series examine particular wars in depth. Wars are defining moments for people and eras. During the twentieth century war became more widespread and terrible than ever before, encouraging new efforts to end war through strategies and organizations of international cooperation and disarmament while also fueling new ideologies and instruments of mass persuasion that fostered distrust and festered old national rivalries. Two world wars during the century redrew the political map, slaughtered or uprooted two generations of people, and introduced and hastened the development of new technologies and weapons of mass destruction. The First World War spelled the end of the old European order and spurred communist revolution in Russia and fascism in Italy, Germany, and elsewhere. The Second World War killed fascism and inspired the final push for freedom from European colonial rule in Asia and Africa. It also led to the Cold War that suffocated much of the world for almost half a century. Large wars begat small ones, and brutal totalitarian regimes cropped up across the globe. After (and in some ways because of) the fall of communism in eastern Europe and the former Soviet Union, wars of competing cultures, national interests, and political systems persisted in the struggle to make a new world order. Continuing, too, has been the belief that military technology can achieve political ends, whether in the superior American firepower that failed to "win" in Vietnam or in the American "smart bombs" and other military wizardry that "won" in the Persian Gulf.

Another theme evident in the series is that throughout the century nationalism has continued to drive events. Whether in the Balkans in 1914 triggering World War I or in the Balkans in the 1990s threatening the post–Cold War peace—or in many other places—nationalist ambitions and forces would not die. The persistence of nationalism is yet another reminder of the many ways that the past becomes prologue.

We thus offer the series as a modern guide to and interpretation of the historic events of the twentieth century and as an invitation to consider how and why those events have defined not only the past and present but also charted the political, social, intellectual, cultural, and economic routes into the next century.

Randall M. Miller
Saint Joseph's University, Philadelphia

Preface

It is often said that there has been greater scientific and technological development in the twentieth century than during any comparable period in human history, and perhaps in all of human history. Nothing demonstrates this transformation of society through science and technology more effectively than the efforts to explore space. Among such endeavors undertaken by various nations, the United States has been a leader almost from the dawn of the space age.

This work is intended as a brief historical synthesis of the contours of space exploration since the first experimentation with rocket technology in the early part of the twentieth century, through the attempts in the 1990s to build a permanently occupied space station. A substantial Chronology of Events outlines the key events in this progression of research and space activity. Thereafter, Chapters 1 through 4 examine several significant issues related to the politics and technology of space exploration. However, they serve only as an introduction to the rich history of this activity.

The chapters elucidate several sets of documents that help to fill out the larger story of the development of space exploration in the last half of the twentieth century. Organized in this volume by theme, they represent the variety of documentary materials available for use by those wishing to understand the rise of space exploration. Also included in this volume are biographical sketches of key people associated with space flight, a listing of the human space flight missions undertaken since 1961, and an annotated bibliography that offers additional reading on this critical subject in the history of the twentieth century.

ACKNOWLEDGMENTS

Whenever historians take on a project of historical synthesis, they stand squarely on the shoulders of earlier investigators and incur a good many intellectual debts. I would like to acknowledge the assistance of several individuals who aided in the preparation of this overview of space exploration history. Lee D. Saegesser, chief archivist in the NASA History Office, was instrumental in obtaining the documents included in the book and in providing other archival services; Stephen J. Garber, assistant historian in the NASA History Office, edited and critiqued the text; Nadine Andreassen helped with proofreading and compilation; the staffs of the NASA Headquarters Library and the Scientific and Technical Information Program provided assistance in locating materials; and archivists at various presidential libraries, at the National Archives and Records Administration, and at other research centers aided with research efforts. Several individuals read portions of the manuscript or talked with me about the project, thereby helping me more than they could ever know. These include Brian Balogh, Donald R. Baucom, Linda Billings, Roger E. Bilstein, Andrew J. Butrica, Michael L. Ciancone, Tom D. Crouch, Robert Dallek, Virginia P. Dawson, Dwayne A. Day, Henry C. Dethloff, David H. DeVorkin, Robert A. Divine, Deborah G. Douglas, Andrew Dunar, Donald C. Elder, Linda Neumann Ezell, Robert H. Ferrell, Michael H. Gorn, Charles J. Gross, R. Cargill Hall, Richard P. Hallion, James R. Hansen, Gregg Herken, Norriss S. Hetherington, Robin Higham, Francis T. Hoban, Joan Hoff, Karl Hufbauer, J. D. Hunley, W. D. Kay, Sylvia K. Kraemer, W. Henry Lambright, John M. Logsdon, Pamela E. Mack, Howard E. McCurdy, John E. Naugle, Allan E. Needell, Lyn Ragsdale, Joseph N. Tatarewicz, Craig B. Waff, Stephen P. Waring, and Mike Wright. All these people would disagree with some of the areas I chose to emphasize, with many of my conclusions, and with a few of my document selections, but such is both the boon and the bane of historical inquiry.

Chronology of Events

1923

December	*Die Rakete zu den Planetenräumen (The Rocket into Interplanetary Space)* by Hermann Oberth was published in Germany; it served to promote considerable discussion of rocket propulsion worldwide.

1926

16 March	In the "Kitty Hawk" of rocketry, Robert H. Goddard launched the world's first liquid-fueled rocket at Auburn, Massachusetts; the rocket traveled 184 feet in 2.5 seconds.

1930

4 April	The American Interplanetary Society was founded in New York City for the "promotion of interest in and experimentation toward interplanetary expeditions and travel." The group was renamed the American Rocket Society (ARS) on 6 April 1934.

1933

14 May	The American Interplanetary Society's rocket No. 2 was successfully fired, attaining a 250-foot altitude in two seconds, at Marine Park on Staten Island, New York.

1935

28 March Robert H. Goddard launched the first rocket equipped with gyroscopic controls near Roswell, New Mexico. The rocket attained an altitude of 4,800 feet, a horizontal distance of 13,000 feet, and a speed of 550 mph.

1936

September Theodore von Kármán, director of the Guggenheim Aeronautical Laboratory at the California Institute of Technology (Caltech) at Pasadena, founded a group that began experiments in design fundamentals of high-altitude sounding rockets; this group later became the Jet Propulsion Laboratory.

1944

September Germany employed the first true ballistic missile, the V-2, against Allied targets in France, Belgium, and England.

1945

August The U.S. Army shipped components for approximately 100 V-2 ballistic missiles from Germany to the White Sands Proving Ground, New Mexico. Aerospace engineer Wernher von Braun and several of his key technical staff working on the V-2 program in Germany during World War II also came to the United States.

26 September The Army's liquid-fueled WAC Corporal rocket was launched to a height of 43.5 miles on its first development flight at White Sands Proving Ground, New Mexico.

1946

15 March The first American-assembled V-2 rocket was static-fired at White Sands Proving Ground.

16 April The first flight test of the American-assembled V-2 rocket was conducted by the Army at White Sands Proving Ground.

24 October V-2 rocket No. 13 launched from White Sands carried a camera that took motion pictures of Earth at an altitude of approximately 65 miles (pictures covered 40,000 square miles).

1949

3 May The Naval Research Laboratory launched the first of twelve Viking rockets from White Sands as part of its sounding rocket research—non-orbital instrument launches exploring the upper atmosphere and testing rocket performance.

1950

1 April The U.S. Army missile staff headed by Wernher von Braun was moved from White Sands Proving Ground in New Mexico to Army Ordnance's Redstone Arsenal in Huntsville, Alabama.

1951

16 January The U.S. Air Force established Project MX-1593 (Project Atlas), a study phase for an intercontinental missile.

1952

18 June Researchers established that heat during re-entry from Earth orbit would be survivable by a craft of blunt shape, which would absorb only one-half of 1 percent of the heat generated by re-entry into the atmosphere.

1953

20 August The Army's Redstone Arsenal launched the first Redstone rocket from its Atlantic Ocean test range at Cape Canaveral, Florida.

1955

2 May The U.S. Air Force approved a plan to inaugurate the Titan intercontinental ballistic missile (SM-68).

26 May The National Security Council, the senior defense policy board in the United States, approved a plan to place into orbit a scientific satellite as part of the nation's involvement in the International Geophysical Year, 1957–1958.

11 June The U.S. Air Force successfully launched the first Atlas intercontinental ballistic missile, and it became operational on 1 September 1959.

9 September To execute the United States' International Geophysical Year commitment, the Department of Defense approved Project

Vanguard to develop a small scientific satellite to go into Earth orbit, managed by the Naval Research Laboratory.

1957

7 August — An Army-JPL (Jet Propulsion Laboratory) Jupiter C rocket fired a scale-model nosecone 1,200 miles down range into the Atlantic Ocean with a summit altitude of 600 miles.

4 October — The Soviet Union launched *Sputnik I*, the world's first artificial satellite, from its rocket testing facility in the desert near Tyuratam in the Kazakh Republic.

3 November — The Soviet Union launched *Sputnik II*, which carried female husky Laika.

17 December — The U.S. Air Force first successfully tested the Atlas intercontinental ballistic missile by launching from its Atlantic complex at Cape Canaveral, Florida.

1958

31 January — As part of Project Explorer, the United States launched its first artificial satellite, *Explorer 1*, atop the interregional ballistic missile *Juno 1*.

15 May — The Soviet Union launched *Sputnik III*, a geophysical laboratory that relayed data about Earth as part of the International Geophysical Year effort.

1 October — The National Aeronautics and Space Administration (NASA) began operating.

6 December — The United States launched *Pioneer 3*, the first U.S. satellite to ascend to an altitude of 63,580 miles.

18 December — An Air Force Atlas booster placed into orbit a communications relay satellite, *Score*, carrying President Eisenhower's Christmas message. This was the first voice sent from space.

1959

2 January — The Soviet Union launched *Luna 1* into solar orbit from its rocket facility at Baikonur. *Luna 1* was the first human-made object to escape Earth's gravity and to be placed in orbit around the sun.

17 February — The United States launched *Vanguard 2*, the first successful launch of a United States principal International Geophysical Year scientific satellite.

28 February — The liquid-hydrogen Thor first stage rocket, and an Agena upper stage rocket, both originally developed by the U.S. Air Force, were used to launch *Discoverer 1*, the first reconnaissance satellite.

3 March — The United States sent *Pioneer 4* to the moon, successfully making the first U.S. lunar flyby.

9 April — After a two-month selection process, NASA unveiled the Mercury astronaut corps.

28 May — The United States launched and recovered two monkeys, Able and Baker, riding in a Jupiter nosecone during a suborbital flight.

12 September — The Soviet Union launched *Luna 2*, which sent back the first clear images of the moon's surface and then crashed on the lunar surface on 14 September.

7 October — The Soviet Union's *Luna 3* lunar probe took pictures of the far side of the moon.

1960

1 April — The United States launched *Tiros 1*, the first successful meteorological satellite observing Earth's weather.

13 April — The United States launched *Transit 1B*, the first experimental orbital navigation system.

1 July — The Army Ballistic Missile Agency of the Redstone Arsenal, Huntsville, Alabama, formally became a part of NASA and was renamed the George C. Marshall Space Flight Center.

12 August — NASA successfully orbited *Echo 1*, a 100-foot inflatable, aluminized balloon for passive communications with a satellite.

1961

12 April — Soviet cosmonaut Yuri A. Gagarin became the first human in space with a one-orbit mission aboard the spacecraft *Vostok 1*.

5 May Astronaut Alan B. Shepard Jr. manned the first U.S. human space flight, *Freedom 7*; it was launched from Cape Canaveral by a Redstone (MR-3) launch vehicle to an altitude of 115 nautical miles and a range of 302 miles.

25 May President John F. Kennedy announced the commitment to land an American on the moon by the end of the 1960s.

21 July The second piloted flight of a Mercury spacecraft took place when astronaut Virgil I. "Gus" Grissom undertook a suborbital mission.

6 August The Soviet Union launched *Vostok 2* with cosmonaut Gherman Titov aboard. Titov completed 17 orbits, the first day-long human space flight mission.

23 August NASA launched *Ranger 1* with the mission of photographing and mapping part of the moon's surface, but it failed to achieve its planned orbit.

19 September NASA administrator James E. Webb announced that the site of the NASA center dedicated to human space flight would be Houston, Texas; at first named the Manned Spacecraft Center, in 1973 it was renamed the Lyndon B. Johnson Space Center.

25 October NASA announced the establishment of the Mississippi Test Facility, a test site for the large Saturn boosters developed for Project Apollo. Located on a bayou in the deep South, it was renamed the John C. Stennis Space Center in 1988.

1962

20 February Astronaut John Glenn became the first American to circle the Earth, making three orbits in the *Friendship 7* Mercury spacecraft.

11–15 August The Soviet Union completed the first long-duration space flight. Cosmonaut Andrian Nicolayev spent four days in space aboard *Vostok 3*.

12 August In the first double flight (occurring at the same time as *Vostok 3* with cosmonaut Nicolayev), the Soviet Union launched *Vostok 4*, with cosmonaut Papel Popovich.

3 October — Astronaut Walter M. "Wally" Schirra Jr. flew six orbits in the Mercury spacecraft *Sigma 7.*

1963

15–16 May — The capstone of Project Mercury took place with the flight of astronaut L. Gordon Cooper, who circled the Earth 22 times in 34 hours aboard the Mercury capsule *Faith 7.*

14–18 June — Soviet cosmonaut Valery Bykovsky set a record aboard *Vostok 5* by orbiting Earth 81 times.

16–19 June — The first woman in space, Soviet cosmonaut Valentina Tereshkova, flew 48 orbits aboard *Vostok 6.*

1964

8 April — The first American Gemini flight took place, an unpiloted test that made four orbits and was successfully recovered.

28 May — The United States placed the first Apollo command module (CM) in orbit in a test flight atop a Saturn I.

28 July — The U.S. *Ranger 7* sent back to Earth 4,300 close-up images of the moon before it impacted on the lunar surface.

1965

18 March — During the Soviet Union's *Voskhod 2* orbital mission, cosmonaut Alexei Leonov performed the first spacewalk, or extravehicular activity (EVA).

23 March — Following two unoccupied test flights, the first operational mission of Project Gemini took place with astronauts Gus Grissom and John W. Young aboard.

6 April — The United States launched *Intelsat I*, the first commercial satellite (communications), into geostationary orbit.

3–7 June — The second piloted Gemini mission, *Gemini IV*, stayed aloft for four days; astronaut Edward H. White II performed the first EVA, or spacewalk, by an American.

14 July — An American space probe, *Mariner 4*, flew within 6,118 miles of Mars after an eight-month journey, providing the first close-up images of the red planet.

21–29 August	During the flight of *Gemini V*, American astronauts Gordon Cooper and Pete Conrad set a record with an eight-day orbital flight.
4–18 December	During the flight of *Gemini VII*, American astronauts Frank Borman and James A. Lovell set a duration record of 14 days in Earth orbit; this record held for five years.
15–16 December	During the flight of *Gemini VI*, U.S. astronauts Wally Schirra and Thomas P. Stafford completed the first true space rendezvous by flying within a few feet of *Gemini VII*.

1966

3 February	The Soviet Union's *Luna 9* soft-landed on the moon in the Ocean of Storms and returned the first photos to Earth from the lunar surface.
1 March	*Venera 3*, built by the Soviet Union, crash-landed on Venus; it was the first human-made object to land on another planet.
16 March	During the flight of *Gemini VIII*, American astronauts Neil A. Armstrong and David Scott docked their spacecraft to an Agena target vehicle, successfully accomplishing the first coupling of two spacecraft.
3 April	The Soviet Union achieved lunar orbit with its *Luna 10* space probe, the first such vehicle to do so.
2 June	*Surveyor 1* landed on the moon and transmitted more than 10,000 high-quality photographs of the surface. This was the first American spacecraft to soft-land on the moon.
3–6 July	During the flight of *Gemini IX*, American astronauts Tom Stafford and Eugene Cernan made a two-hour EVA.
18–21 July	During the flight of *Gemini X*, American astronauts Mike Collins and John Young made two rendezvous and docking maneuvers with Agena target vehicles; they also completed a complex EVA.
10 August 1966–1 August 1967	The Lunar Orbiter project was conducted for a full year as a means of mapping the surface of the moon in preparation for the Apollo landings.

11–15 November	The last Gemini flight, *Gemini XII*, was launched with astronauts Jim Lovell and Edwin E. "Buzz" Aldrin Jr., who undertook three EVAs and a docking with an Agena target vehicle.
1967	
27 January	At 6:31 P.M., during a simulation aboard *Apollo-Saturn (AS) 204* on the launch pad at Kennedy Space Center, Florida, after several hours of work, a flash fire broke out in the pure oxygen atmosphere of the capsule. Flames engulfed the capsule, and the three astronauts aboard—Gus Grissom, Roger Chaffee, and Edward White—died of asphyxiation.
24 April	During the return of *Soyuz 1*, Soviet cosmonaut Vladimir Komarov died when the capsule returned to Earth in a spin.
9 November	During the flight of *Apollo 4*, an unpiloted test of the launcher and spacecraft, NASA proved that the combination could safely reach the moon.
1968	
22 January	In *Apollo 5*, NASA made the first flight test of the propulsion systems of the lunar module's ascent/descent capability.
14 September	In a significant first, the Soviet Union sent its *Zond 5* lunar mission capsule on an unpiloted flight around the moon and brought it back safely to Earth.
11–12 October	The first piloted flight of the Apollo spacecraft, *Apollo 7*, and Saturn IB launch vehicle involved astronauts Wally Schirra, Donn F. Eisele, and Walter Cunningham, who tested hardware while in Earth orbit.
26 October	The Soviet Union's *Soyuz 3* made a rendezvous with *Soyuz 2* in a mission that repeated the basic flights of *Gemini VI* and *Gemini VII*.
21–27 December	*Apollo 8* made the first circumlunar flight (around the moon) with a crew aboard: Frank Borman, James A. Lovell Jr., and William A. Anders.

1969

16–24 July The first lunar landing mission, *Apollo 11*, achieved the late President Kennedy's goal of landing an American on the moon before the end of the decade; astronauts Neil A. Armstrong, Buzz Aldrin, and Michael Collins were aboard.

15 September The presidentially appointed Space Task Group issued its report on the post-Apollo space program, calling for a far-reaching space program that included development of a space station, a reusable space shuttle, a moon base, and a manned expedition to Mars.

1970

11–17 April The flight of *Apollo 13* became one of the near-disasters of the Apollo program when an oxygen tank in the *Apollo* service module exploded and the crew worked with ground controllers to find a way safely back to Earth.

1–18 June In the Soviet Union's *Soyuz 9*, cosmonauts Nikolayev and Sevastyanov set an 18-day endurance record that broke the *Gemini VII* record of 1965.

12 September Failing to win the race to place a human on the moon, the Soviet Union succeeded in returning to Earth lunar samples from a robotic probe, *Luna 16*.

10 November The Soviet Union launched *Luna 17*, a robotic probe to the moon that carried *Lunokhod 1*—a small moon rover that operated under remote control for several months.

1971

19 April The Soviet Union launched the world's first space station, *Salyut 1*, and a crew deployed to it on 23 April 1971.

26 July–7 August The first of the longer, expedition-style lunar landing missions, *Apollo 15*, was the first to include the lunar rover to extend the range of the astronauts on the moon, allowing them to bring back 173 pounds of moon rocks.

1972

5 January	President Richard M. Nixon announced the decision to develop a space shuttle, which was first flown in space on 12–14 April 1981.
3 March 1972–Present	To prepare the way for a possible mission to the four giant planets of the outer solar system, *Pioneer 10* and *Pioneer 11* went to Jupiter and Saturn and, from there, outside the solar system.
23 July 1972–Present	*Landsat 1*, the first of a series that would operate through the end of the century, was launched from Kennedy Space Center to carry out an Earth resource mapping mission that provided data on vegetation, insect infestations, crop growth, and associated land-use information.
7–19 December	*Apollo 17* was the last of the six Apollo missions to the moon, and the only one to include a scientist—astronaut/geologist Harrison Schmitt—as a member of the crew.

1973

25 May 1973–February 1974	NASA conducted research on the orbital workshop *Skylab*, the first U.S. attempt to build a space station.

1975

15–24 July	NASA conducted the Apollo-Soyuz Test Project as an international docking mission between the United States and the Soviet Union, demonstrating graphically the détente between the two Cold War superpowers.
20 August 1975–21 May 1983	NASA conducted the Viking soft-landing mission to Mars, landing two probes on the surface that successfully revolutionized scientific understanding of the red planet.

1976

22 June	The Soviet Union launched *Salyut 5*, a military space station.
20 July	The *Viking 1* planetary lander touched down on the Chryse Planitia (Golden Plains) of Mars after a voyage of nearly one year.

1977

18 February — The first orbiter, *Enterprise* (OV-101), was first flown in flight tests atop Boeing 747 ferrying aircraft at NASA's Dryden Flight Research Facility in southern California.

12 August — The *Enterprise* had its first free flight test at NASA's Dryden Flight Research Facility in the high desert of southern California at Muroc Dry Lake.

20 August 1977–Present — NASA undertook the Voyager program, with two probes, to the outermost giant planets, thereby greatly expanding knowledge of the outer solar system.

29 September — The Soviet Union launched *Salyut 6*, a civilian space station that remained operational for three and a half years. The last mission to it was *Soyuz T-4*, launched on 12 March 1981. During active life, *Salyut 6* was home for 16 crews and was occupied for 676 days.

1978

20 May 1978–9 May 1979 — The United States undertook a mission to Venus, *Pioneer Venus 1* and *2*, that relayed scientific data about climate, chemical makeup, and atmospheric conditions of the planet.

1979

11 July — *Skylab* finally impacted the Earth's surface after its orbit had been in decay for more than two years; debris dispersed from the southeastern Indian Ocean across a sparsely populated section of western Australia.

1981

12–14 April — Astronauts John W. Young and Robert L. Crippin flew the space shuttle *Columbia* on the first flight of the Space Transportation System (STS-1).

1983

18–24 June — The *Challenger* (STS-7) made history by having as a member of the crew scientist-astronaut Sally K. Ride, the first American woman in space.

30 August	The *Challenger* (STS-8) made history by having as a member of its crew the first black American astronaut, Guion S. Bluford.

1984

25 January	President Ronald Reagan announced the decision to build a space station within a decade.
3–10 February	During the flight of STS-41B, the *Challenger*, astronauts performed extravehicular activity using the Manned Maneuvering Unit (MMU).
15 December 1984–14 March 1986	An international armada of spacecraft encountered the Comet Halley during its closest approach to Earth in 76 years.

1985

3–7 October	In the first U.S. Department of Defense–dedicated mission, the *Atlantis* (STS-51J) deployed a classified satellite.

1986

28 January	In the worst space disaster to date, the *Challenger* (STS-51L) exploded and its crew of seven was killed 73 seconds after launch. The seven crew members were Francis R. (Dick) Scobee, Michael J. Smith, Judith A. Resnik, Ronald E. McNair, Ellison S. Onizuka, Gregory B. Jarvis, and Sharon Christa McAuliffe. The Soviet Union launched its *Mir* space station, which would later be incorporated into the international space station effort.
May	The National Commission on Space, chaired by Thomas O. Paine, issued its report on the U.S. civil space program: *Pioneering the Space Frontier: An Exciting Vision of Our Next Fifty Years in Space*. The report advocated an aggressive space effort oriented toward the exploration and eventual colonization of the moon and the other planets of the solar system.
12 May	James C. Fletcher became the NASA administrator for a second time, with the goal of reforming NASA after the January 1986 *Challenger* accident.

6 June — The *Report of the Presidential Commission on the Space Shuttle* Challenger *Accident* identified problems with the O-rings on the solid rocket boosters as the cause of the fatal explosion.

8 July — NASA created the Office of Safety, Reliability, Maintainability, and Quality Assurance in response to findings from the teams investigating the *Challenger* accident.

15 August — President Ronald Reagan announced that NASA would no longer launch commercial satellites, except those that were shuttle-unique or had national security or foreign policy implications.

15 August — NASA secured presidential and congressional support for the acquisition of a replacement orbiter for *Challenger*, enabling the agency to continue its efforts to build an international space station.

1988

29 September–3 October — The space shuttle returned to flight following the *Challenger* accident with the launch of *Discovery*.

21 December — Two Soviet cosmonauts returned to Earth following a record-setting period in space aboard the Soviet space station *Mir*. Cosmonauts Titov and Manarov far outdistanced any other spacefarers after a period of one year, 22 hours, and 39 minutes in space.

1989

5 May 1989–1993 — The highly successful *Magellan* radar mapping mission to Venus took place.

25 August — *Voyager 2*, operated by NASA's Jet Propulsion Laboratory, encountered Neptune within 3,115 miles and then moved on to encounter the moon Triton.

19 October 1989–Present — The *Galileo* spacecraft was launched from STS-34 and began a gravity-assisted journey to Jupiter, arriving in December 1995. This began a two-year encounter with the planet in which *Galileo* sent back to Earth scientific data about the density and chemical makeup of the giant planet's cloud cover.

1990

24 April 1990–Present — The Hubble Space Telescope was deployed from a space shuttle in Earth orbit. The instrument proved faulty, and had to be repaired in orbit in December 1993.

17 December — The presidentially chartered Advisory Committee on the Future of the U.S. Space Program issued a report advocating human space flight, robotic probes, space science, applications, and exploration within a tightly constrained budget.

1992

25 September 1992–29 October 1993 — The *Mars Observer* satellite mission took place, although it was lost as it neared the red planet.

6 October — NASA administrator Daniel S. Goldin and Russian Space Agency director Yuri Koptev signed two cooperative agreements in Moscow regarding human space flight, including participation in an international consortium to build a space station, with the United States as the senior partner.

1993

2–12 December — Astronauts aboard the *Endeavour* (STS-61) executed a highly successful mission to repair the optics of the Hubble Space Telescope (HST) and perform routine servicing on the orbiting observatory, conducting five extravehicular activities.

1994

3–11 February — The *Discovery* (STS-60) flew a historic mission with a Russian cosmonaut to rendezvous with the *Mir* space station as part of the international effort to build a space station.

1995

3–11 February — Exactly one year after a cooperative flight with the Russians in STS-60, NASA's *Discovery* flew by the Russian space station *Mir* under the control of the first woman pilot, Eileen M. Collins.

27 June–7 July — One of the most significant missions to take place in recent years occurred when American astronaut Norman E. Thagard, a physician, spent more than three months on the Russian space station *Mir* and then was returned to Earth on the space shuttle *Atlantis* after it completed the first shuttle/*Mir* docking mission.

11–20 November

This mission by the space shuttle *Atlantis* carried up and attached a Russian-built docking port and orbiter docking system to the *Mir* space station for use in future shuttle dockings.

1996

22–31 March

In this *Atlantis* shuttle mission to dock with the Russian space station *Mir*, the United States left astronaut Shannon Lucid, the first U.S. woman to fly on the station, aboard for a total of five months.

16–26 September

The *Atlantis* docked with *Mir* and retrieved Shannon Lucid and left John Blaha for continued joint operations aboard the Russian station.

1997

11–21 February

In a record five extravehicular activity (EVA) operations, astronauts from the shuttle *Discovery* performed the second Hubble Space Telescope servicing mission. This mission replaced the near-infra red camera (NICMOS), the two-dimensional spectrograph, and repaired insulation on the telescope.

25 June

During the attempted docking of the Russian resupply vessel, *Progress*, with the Russian space station *Mir* the vessel collided with the science module, *Spektor*, attached to *Mir*. The module decompressed and its solar arrays were knocked out of service. Although the crew of two Russian cosmonauts and one American astronaut, Michael Foale, are uninjured, the accident crippled the space station and led to a series of crises in space. The Russian Space Agency managed to keep the station operational until it could be resupplied and repaired.

4 July

The inexpensive *Mars Pathfinder* (costing only $267 million) landed on Mars, after its launch in December 1996. A small, 23-pound robotic rover, named *Sojourner*, departed the main lander and began to record weather patterns, atmospheric opacity, and the chemical composition of rocks washed down into the Ares Vallis flood plain, an ancient outflow channel in Mars' northern hemisphere. This vehicle completed its projected milestone 30-day mission on 3 August 1997, capturing far more data on the atmosphere, weather, and geology of Mars than scientists had expected. In all, the *Pathfinder* mission returned more than 1.2 gigabits (1.2 billion bits) of data and over 10,000 tantalizing pictures of the Martian landscape.

25 September–6 October

In this seventh docking mission with the Russian space station *Mir*, the shuttle *Atlantis* delivered three Russian air tanks and nine *Mir* batteries (170 pounds each). They also delivered a *Spektor* module repair kit (500 pounds), which enabled the station crew to begin seriously needed repairs from the *Progress* collision of June 25. The mission also delivered 1,400 pounds of water, 1,033 pounds of U.S. science items, and 3,000 pounds of Russian supplies. During this mission Russian cosmonauts Parazynski and Titov conduct an EVA to retrieve four environmental effects space exposure experiments (MEEPS) on *Mir*'s module. *Atlantis* also flew around *Mir* to assess the damage to the station. The astronaut Michael Foale also departed for Earth after a stay of nearly five months and was replaced by astronaut David Wolf.

SPACE EXPLORATION EXPLAINED

1

Historical Overview of Space Exploration

Throughout history, humans have expressed a fascination with the natural universe and a desire, translated into action, to learn more about it. Ancient peoples such as the Greeks and Romans, as well as others in the non-Western tradition, have expended enormous resources to understand the places and phenomena beyond their immediate lines of sight. In this overview we will examine the historical background of exploration and settlement, and we will place it in the context of the three great ages of discovery in Western civilization since the Renaissance. In each of the three ages, major new developments made possible subsequent periods of discovery. Space flight is a part of the last, or most recent, age of exploration.

THE THREE GREAT AGES OF EXPLORATION

There have been three great ages of exploration in the history of Western civilization, the last of which informs our everyday lives even as it remains in its infancy. Space exploration, which is just one attribute of this third age, helps define it. Each age of exploration became central to the worldview of Western civilization. In other times and places, exploration had occurred and civilizations had experienced challenges of new lands and situations similar to those of the three great ages; but those other instances were markedly different in that exploration did not become the basis for a worldview.

In each of the three great ages of exploration in Western civilization, five critical factors shaped the nature of exploration. Without any one factor, it is unlikely the age would have developed as it did. The first factor involves the political will to carry out expeditions of exploration. Since most expeditions in all three ages—and virtually all the large expeditions—have been sponsored by governments, high-level decisionmakers had to agree that the expenditure of funds for exploration was in the best interest of the state. Without that political will, discovery and exploration could not take place. Second, the sponsoring organization must have a sufficiently productive and stable economic base to cover spending a considerable percentage of its budget in this manner. Third, the citizens or subjects of the sponsoring entity must agree with, or acquiesce to, the leadership's decision that exploration of the unknown is acceptable and worthy of support. Without this tacit approval, leaders cannot sustain exploration for long. Fourth, there has to be a scientific basis of knowledge to which information gathered in exploration might be added. Without such an anticipated return, there can be little continuation of political will or popular support. Finally, enabling technologies must be available to allow the explorers to succeed.

The first great age of discovery began during the European Renaissance of the fifteenth century. Thereafter nothing would ever be the same again, as western Europe was transformed by contact with new lands and peoples during voyages of discovery between the fifteenth and eighteenth centuries. During this era, mariners from the great seafaring nations of western Europe essentially redrew both the map and the conception of the world. When they were done, the contours of Earth's great continents had been approximated and the general size and shape of the physical world had been determined with relative accuracy. As scientist Peter Martyr wrote in 1493, just as this age of exploration was beginning, "Enough for us that the hidden half of the globe is brought to light. . . . Thus shores unknown will soon become accessible; for one in emulation of another sets forth in labours and mighty perils."[1] Exploration during the first age of discovery took place mainly over the oceans, as European sailors mapped the coastlines of the Americas, Africa, Australia, and even Antarctica. The great explorers included Christopher Columbus, Ferdinand Magellan, Henry Hudson, Jacques Cartier, and James Cook, among others. Indeed, the classic expression of this age of exploration was the circumnavigation of the globe. The explorers' reconnaissances were critical to the success of the age, and the bulk of their accumulated data on peoples and geographies transformed Western civilization.

Exploration during the second age (which began before the end of the first and even coincided with it in certain places and circumstances) took place predominantly overland, as European adventurers filled in many details of the continental interiors. In the process, geographical knowledge continued to increase, but so did information about peoples and natural history. The expedition of Coronado in 1540 to southwestern North America, the Lewis and Clark expedition to the Rocky Mountain west of North America in 1804–1806, the efforts of Sir Richard Burton and Stanley and Livingston in Africa, and travels to the sources of the Amazon in South America and the Nile in Africa—all characterized the second age of exploration. This age ended with the conclusion of the last great expeditions into the interiors of the continents in the later nineteenth century. It, too, led to a massive accumulation of data about lands new to European civilization, and it transformed the scientific world with the cataloging of much new information.

The third great age of exploration, fundamentally a twentieth-century phenomenon, is strikingly different in terms of the areas investigated. This age has been characterized by the movement of explorers, both human and robotic, into realms where humans cannot live without the benefit of artificial apparatus. Exploration on the two poles of Earth, under the oceans, and in space all suggest a new frontier of discovery and inquiry. The explorers of this new age include Richard Byrd with his epic North Pole flight of 1926, Jacques Cousteau and the voyages of his scientific vessel *Calypso*, Soviet cosmonaut Yuri Gagarin as the first human in space in 1961, and American astronaut Neil Armstrong walking on the moon in 1969.

What was true in the earlier ages of exploration was especially so in this later period: huge investments in technology over long periods made possible the explorations. For example, it cost approximately $25 billion to mount the Apollo program to explore the moon in the 1960s. Earlier expeditions may have been smaller, but they were proportionately expensive in terms of their era. The returns on investment in this third age of exploration, which are only now beginning to be realized, involve the geophysical inventory of a planet and the exploitation of these new regions for all types of commercial ventures that have changed our lives. For example, remote sensing satellites have made life strikingly different from what it was only a generation ago as satellite images of weather patterns enable meteorologists to forecast storms, as communications satellites transform our ability to move information, and as global positioning satellites instantaneously provide reliable geographical information.

The sum total of this transformation has informed our perspective on the world. It has also underscored the unique role of exploration by civilizations. As it became ingrained in Western culture beginning in the sixteenth century, exploration gave birth to the scientific revolution and the transformation of Western beliefs and ideals.

THE RATIONALE FOR SPACE EXPLORATION

Why spend staggering sums to carry out a space exploration program? There have been essentially three priorities. The first involved Cold War rivalries between the United States and the Soviet Union, and the desire to demonstrate the technological superiority of one form of government over another: a democratic state versus a communist dictatorship. The second priority was the lure of discovery of the unknown. The third was adventure. The first priority, oriented toward national security, governed much of the effort in the first three decades of the space age, but it has ceased to be important in the post–Cold War era of the 1990s. The second and third priorities, however, continue to motivate those who believe in human exploration; these priorities are just as attractive today as they were some forty years ago at the birth of the space age.

With the end of the Cold War between the United States and the Soviet Union, advocates of an aggressive space exploration program faced the loss of the first priority (national security) and had to refocus on the others. Forty years ago, before the *Sputnik* crisis,[2] space enthusiasts were motivated by an expansive view of human voyages of discovery, exploration and settlement of the moon and other planets of the solar system, and eventual interstellar travel. The sense that humans have to "get off this planet" if the human race is to survive indefinitely is a compelling aspect of the dream of space flight. In this view the human component of space flight has been central; robotic probes and applications satellites have been a useful but decidedly less important aspect of the space exploration agenda. But adventure and discovery, as well as the long-range goals of exploration and colonization, have never in and of themselves seemed important enough for political leaders to justify the ventures space advocates have proposed.

There seems little doubt but that adventure and discovery, the promise of exploration and colonization, were the motivating forces behind the small cadre of space program advocates in both the United States and the Soviet Union prior to the 1950s. Most American advocates of aggressive space exploration efforts invoked an extension of the popular notion of the frontier, with its positive images of territorial discovery, scientific discovery, exploration, colonization, and use.

There were many ways by which Americans and Soviets became aware that flight into space was a possibility, ranging from (1) science fiction literature and film that were becoming more closely tied to reality than ever before, to (2) speculations by science fiction writers about possibilities already being made real, to (3) serious discussions of the subject in respected popular magazines. In the United States, among the most important serious efforts were those of the handsome German emigré Wernher von Braun, who worked for the U.S. Army at Huntsville, Alabama. In addition to being a superbly effective technological entrepreneur, von Braun managed to captivate the powerful print and electronic communications media in the early 1950s; for the next two decades, no one was a more effective promoter of space exploration. In the Soviet Union, public advocates in the sciences recommended space flight efforts to advance the Marxist cause, the completion of a utopian order of civilization worldwide.

At the same time that space exploration advocates—both buffs and scientists—were generating an image of space flight as genuine possibility and no longer fantasy, and proposing how to accomplish a far-reaching program of lunar and planetary exploration, another critical element entered the picture: the role of space flight in national defense and international relations. Space partisans readily hitched their exploration vision to the political requirements of the Cold War—in particular, to the belief that whichever nation occupied the "high ground" of space would dominate the territories underneath it.

THE SPACE RACE

The Cold War rivalry between the United States and the Soviet Union was the key that opened the door to aggressive space exploration on both sides—not as an end in itself, but as a means to achieving technological superiority in the eyes of the world over an adversary. From the perspective of the 1990s, it is difficult to appreciate the almost hysterical concern over nuclear attack that preoccupied Americans in the 1950s, and the Soviets' fears of aggression from the powerful United States. Both nations became preoccupied with fear of death by nuclear war. Schoolchildren were required to practice civil defense techniques and shield themselves from nuclear blasts, in some cases simply by crawling under their desks. Communities practiced civil defense drills, and families built bomb shelters in their backyards. In the popular culture, nuclear attack was inexorably linked to the space above the United States and the Soviet Union, from which the dreaded attack would come.

The perception of space as the "high ground" of Cold War competition gained credibility from the atomic holocaust literature of the era. The danger of surprise attacks had been burned into America's national consciousness by the Japanese attack on Pearl Harbor in 1941 and into the Soviets' by the devastation of their nation by Hitler's invasion in 1941 and four years of war. Combine this understanding of terror with the proliferation of American and Soviet nuclear weapons, and the nightmare had become reality.

After an arms race with its nuclear component and a series of hot and cold crises in the Eisenhower era, with the launching of *Sputnik* in 1957 the threat of holocaust felt by most Americans and Soviets was now not just a possibility, but a probability. In 1957–1958, aggressive space exploration emerged directly out of the Cold War rivalries of the United States and the Soviet Union. In the contest over the ideologies and allegiances of the world's nonaligned nations, space exploration became contested ground. The Soviets gained the upper hand in this competition in 1957 with *Sputnik I*, the first artificial satellite to orbit the Earth as part of a larger scientific effort associated with the International Geophysical Year.

Even while U.S. officials congratulated the Soviet Union for this accomplishment, many Americans thought that the Soviet Union had staged a tremendous coup for the communist system at U.S. expense. Because of this perception, Congress passed and President Dwight D. Eisenhower signed the National Aeronautics and Space Act of 1958. This legislation established the National Aeronautics and Space Administration (NASA) with a broad mandate to explore and use space for the benefit "of all mankind."

The Soviet Union, while not creating a separate organization dedicated to space exploration, infused money into its various rocket design bureaus and scientific research institutions. The chief beneficiaries of Soviet space flight enthusiasm were the design bureau of Sergei P. Korolev, the chief designer of the first Soviet rockets used for the Sputnik program; and the Soviet Academy of Sciences, which devised experiments and built the instruments that were launched into orbit. With huge investments made in space flight technology by Soviet premier Nikita Khrushchev, the Soviet Union accomplished one public relations coup after another against the United States during the late 1950s and early 1960s.

In the United States, within a short time after NASA's formal organization the new agency also took over management of space exploration projects from other federal agencies and began to conduct space science missions—such as Project Ranger to send probes to the moon, Project Echo to test the possibility of satellite communications, and Project Mercury to ascertain the possibilities of human space flight. Even so, these activities were constrained by a modest budget and a measured pace on the part of NASA leadership.

THE RACE TO THE MOON

Those constraints were suddenly lifted in 1961 when President John F. Kennedy, responding to perceived challenges to U.S. leadership in science and technology, announced a lunar landing effort that would place an American on the moon before the end of the decade. Kennedy unveiled this commitment, called Project Apollo, before Congress on 25 May 1961 in a speech on "Urgent National Needs," billed as a second State of the Union message.

For the next eleven years, NASA was consumed with carrying out the Kennedy mandate. This effort required tremendous expenditure, more than $20 billion over the life of the program, to make it a reality by 1969.

Only the building of the Panama Canal rivaled the Apollo program's size as the largest nonmilitary technological endeavor ever undertaken by the United States; only the Manhattan Project to build an atomic bomb in World War II was comparable in a wartime setting. The human space flight imperative was a direct outgrowth of Kennedy's decision; Projects Mercury (in its latter stages), Gemini, and Apollo were each designed to execute it.

NASA eventually landed six sets of astronauts on the moon between 1969 and 1972. The first landing mission, *Apollo 11*, succeeded on 20 July 1969, when astronaut Neil Armstrong first set foot on the lunar surface, telling millions of listeners that it was "one small step for [a] man—one giant leap for mankind." Five more landing missions followed *Apollo 11* at approximately six-month intervals through December 1972.

The Soviet Union also engaged in its own program for a landing on the moon, although for years its leaders denied the existence of such an effort (to avoid embarrassment if it proved unsuccessful). Korolev and the leaders of other rocket design bureaus spent billions of rubles, built space hardware including a lunar lander and the N-1 moon rocket, and warred among themselves for patronage from Khrushchev and his successors in the Politburo. Korolev died unexpectedly in 1966, sending awry whatever political stability had been present in the Soviet lunar program. Its efforts quickly unraveled. The Americans bested the Soviets in the moon race, and Russia's communist leaders buried any evidence of the aborted project until the fall of the communist regime in 1989. "Korolev's achievement, Khrushchev's use of it, and the response to both by the Americans defined the politics of technology through two decades," wrote Pulitzer Prize–winning historian Walter A. McDougall in 1985. "And even though the Soviet bear ran in circles for a time, the scientists of Korolev's generation, struggling within, while working for, the Soviet system, laid the foundation for a postindustrial communist Superpower second to none."[3]

SUSTAINED SPACE EXPLORATION

NASA went into a holding pattern after the completion of Project Apollo. After measured but unsustained success with an orbital workshop, *Skylab*, NASA's major program for the 1970s was the development of a reusable space shuttle that would travel back and forth between Earth and space more routinely and economically than ever before. In 1981 the first operational orbiter, *Columbia*, took off from the Kennedy Space Center in Florida. Five years later there had been 24 shuttle flights, but during the launch of *Challenger* on 28 January 1986 a leak in the joints of one of two solid rocket boosters detonated the main liquid fuel tank. Seven astronauts died in this accident, the worst in the history of space flight. The explosion became one of the most significant events of the 1980s, as billions of people around the world saw the accident on television and empathized with any one or more of the crewmembers killed. Following the accident, the shuttle program went into a two-year hiatus while NASA worked to redesign the system and revamp its management structure. The space shuttle finally returned to flight without further incident on 29 September 1988.

In addition to those important human space flight programs, there were significant scientific probes to the moon and planets, as well as Earth-observing systems placed in orbit. Among the most significant have been the Viking missions to Mars, the culmination of a series of missions to explore the planet that began in 1964 with *Mariner 4* and continued with other missions. Viking consisted of two spacecraft designed to orbit Mars and to land and operate on the planet's surface. Launched in 1975, the probes spent nearly a year cruising to Mars and landed during America's bicentennial year, 1976.

Another important probe was the Voyager mission to the outer solar system. Launched in 1977, these spacecraft visited their primary targets of Jupiter and Saturn and then went on to all the giant outer planets. The two Voyagers took well over 100,000 images of the outer planets, rings, and satellites, as well as millions of magnetic, chemical spectra, and radiation measurements. Without question, they returned information to Earth that revolutionized the science of planetary astronomy.

More recently the Hubble Space Telescope, although initially impaired, has returned exceptional scientific data about the origins and development of the universe; the Magellan mission has radar-imaged Venus; and the Galileo probe to Jupiter, although experiencing problems with its communication system, has returned important data about the fifth planet. At the same time, because of budgetary and other constraints, NASA leaders have moved toward the building of a large number and variety of small, inexpensive satellites rather than a few large, expensive spacecraft. In the 1990s the NASA administration began to urge a new philosophy of "smaller, cheaper, faster" for the agency's space probes and advocated a mixture of large and small spacecraft to avoid the long hiatus that would occur if a mission failed.

The Soviet Union was not inactive following the race to the moon. It made a special province of its program to advance sustained human operations in Earth-orbit and launched a series of space stations by which it conducted an expansive set of scientific experiments. In 1971 the Soviet Union launched its heralded *Salyut* space station on the tenth anniversary of the flight of the first human in space. In 1973 it launched more stations, but they failed to achieve orbit. In 1977 it launched *Salyut VI*, which enjoyed a long service life and routinely maintained cosmonauts in orbit for up to six months at a time. In 1986 the Soviet Union launched *Mir*, a relatively advanced space station that by the mid-1990s had become a critical component in the multinational effort led by the United States to establish a permanent presence in space. Except for a few months during the collapse of the Soviet empire and its replacement by a coalition of market economy states in 1989, from 1986 to 1997 *Mir* has been permanently occupied. While it has been plagued by recurrent problems since the spring of 1997, its scientific experiments have added enormously to understanding about the effects of space on the human body.

The Soviet Union also engaged in a dynamic planetary science program during the 1970s and 1980s. In September 1970 it soft-landed a probe on the moon and returned rock and soil samples. It followed this with nine more successful robotic missions to the moon through 1976. Nearly cornering the market on science missions to Venus, the Soviet Union also sent ten Venera probes to the planet between 1970 and 1983. The Soviets also sent probes to Mars, orbited scientific and applications spacecraft around Earth, and explored asteroids and Halley's Comet using robotic probes during the same era.

Just as the Cold War was the driving force behind big budgets for space exploration in the 1960s and continued to influence expenditures in the 1970s and 1980s, its end has been a critical component in the search for a new space policy for both the United States and Russia in the 1990s. What did offer promise, however, was a different outlook on international relations, that is, cooperation with international partners. Whereas Apollo in the United States and the N-1 moon rocket effort in the Soviet Union had been entirely national projects because of their Cold War origins, in post–lunar landing activities there were opportunities for greater partnership with foreign allies.

In terms of the effort to promote an international space station during the 1980s, from the outset both NASA and the Reagan administration emphasized it as an *international* effort. NASA forged agreements among thirteen nations—including Japan, Canada, and several European nations—to take part in the Space Station *Freedom* program in 1985. This effort was intended to maximize technological capability while reducing the cost to each participating nation. But the ingenious aspect of the partnership arrangement was that it helped replace the rivalry of the Cold War as a means of stabilizing support and funding for NASA by linking the space program's major effort to yet another U.S. foreign policy objective—in this case, international cooperation. In so doing, every partnership brought greater legitimacy to the overall program and helped insulate it from drastic budgetary and political changes.

This new spin on an old idea took an even more significant turn in the fall of 1993 when NASA negotiated a landmark decision to include Russia in the building of an international space station. In the post–Cold War era, as the United States wrestles with foreign policy questions aimed at supporting democratic reforms in eastern Europe and Russia, this decision has so far provided an important linkage for continuation of the space station effort. The decision has come at a time when many politicians are convinced that other reasons such as cost, technological challenge, and return on investment make continued space research an uninviting endeavor. Just as some members of Congress in the early 1960s were uninterested in space exploration except as a tool of foreign policy goals, some members today consider the modern cooperative agreements for joint space exploration more important than other issues associated with the effort.

The combination of technological and scientific advancement, political competition with the Soviet Union, and changes in popular opinion about space flight came together in a focused way in the 1950s to affect public policy in favor of an aggressive space program. This found tangible expression in efforts of the 1950s and 1960s to move forward with an expansive space program and the budgets necessary to support it.

CONCLUSION

A longstanding fascination with discovery and investigation, as well as the revered frontier experience, have nourished much of the interest by the peoples of the United States in space flight during the third great age of exploration. The compelling nature of these perceptions has ensured the success of the nation's endeavors in space thus far.

NOTES

1. Quoted in Daniel Boorstin, *The Discoverers* (New York: Random House, 1983), p. 145.
2. The Sputnik crisis was precipitated by the 4 October 1957, launch of an Earth satellite by the Soviet Union, thus bringing a strong U.S. reaction that portended important Cold War concerns.
3. Walter A. McDougall, *". . . The Heavens and the Earth": A Political History of the Space Age* (New York: Basic Books, 1985), p. 297.

2

The Race to the Moon

On 25 May 1961, President John F. Kennedy announced to the nation a goal of sending an American safely to the moon and back before the end of the decade. This decision involved much study and review prior to making it public, and tremendous expenditure and effort to make it a reality by 1969. Indeed, the Apollo program became the largest nonmilitary technological endeavor ever undertaken by the United States. The human space flight imperative was a direct outgrowth of the call. Projects Mercury (in its latter stages), Gemini, and Apollo were each designed to execute it. The mission was finally accomplished on 20 July 1969, when *Apollo 11*'s astronaut Neil Armstrong stepped out of the lunar module and set foot on the moon.

A COLD WAR CHALLENGE

Had the balance of power and prestige between the United States and the Soviet Union remained stable in the spring of 1961, it is quite possible that Kennedy would never have advanced his moon program and the direction of American space efforts might have taken a radically different course. Kennedy seemed quite happy to allow NASA to execute Project Mercury at a deliberate pace, working toward the orbiting of an astronaut sometime in the middle of the decade, and to build on the satellite programs that were yielding excellent results in terms of both scientific knowledge and practical applications. Jerome Wiesner reflected, "If Kennedy could have opted out of a big space program without hurting the country in his judgment, he would have."[1]

But Kennedy did not believe he could opt out after Soviet cosmonaut Yuri Gagarin became the first human to orbit in space on 12 April 1961 aboard *Vostok 1.* The great success of that feat made the gregarious Gagarin a global hero, and he was an effective spokesman for the Soviet Union until his death in 1968 from an aircraft accident. It was only a salve on an open wound, therefore, when Alan Shepard became the first American in space during a 15-minute suborbital flight on 5 May 1961 by riding a Redstone booster in his *Freedom 7* Mercury spacecraft.

Comparisons between the Soviet and American flights were inevitable thereafter. Gagarin had orbited the Earth; Shepard had been a cannonball shot from a gun. Gagarin's *Vostok* spacecraft had weighed 10,428 pounds; *Freedom 7* weighed 2,100 pounds. Gagarin had been weightless for 89 minutes; Shepard for only 5 minutes. "Even though the United States is still the strongest military power and leads in many aspects of the space race," wrote journalist Hanson Baldwin in the *New York Times* not long after Gagarin's flight, "the world—impressed by the spectacular Soviet firsts—believes we lag militarily and technologically."[2]

By any measure, the United States had not demonstrated technical equality with the Soviet Union, and that fact worried national leaders because of what it would mean in the larger Cold War environment. This and a host of international problems—the aborted Bay of Pigs invasion of Cuba, an ill-fated Geneva summit, and a crisis over Germany that led to the construction of the Berlin Wall—also played a role in Kennedy's decision.

President Kennedy unveiled the commitment to execute Project Apollo on 25 May 1961. He told the American people:

> I believe this Nation should commitment itself to achieving the goal, before this decade is out, of landing a man on the moon and returning him safely to earth. No single space project in this period will be more impressive to mankind, or more important for the long-range exploration of space; and none will be so difficult or expensive to accomplish.[3]

In the end, a unique confluence of political necessity, personal commitment and activism, scientific and technological capability, economic prosperity, and public mood made possible the 1961 decision to carry out a forward-looking lunar landing program. A complex system of ties between various people, institutions, and interests prompted the Apollo decision, but it fell to NASA and other organizations of the federal government to accomplish the task set out in a few short paragraphs by President Kennedy.

GEARING UP FOR PROJECT APOLLO

The first challenge in meeting the presidential mandate was to secure funding. Congress enthusiastically appropriated funding for Apollo immediately after the president's announcement, and the space agency's annual budget increased from $500 million in 1960 to a peak of $5.2 billion in 1965—about 5.3 percent of the total federal budget. In all, it was an enormous undertaking, costing $25.4 billion (about $95 billion in 1990 dollars). This money was used to hire workers, expanding the agency's civil service payroll to over 36,000 people by 1965 from the 10,000 employed at NASA in 1960. Additionally, NASA's contractor work force increased by a factor of ten, from 36,500 in 1960 to 376,700 in 1965. Private industry, research institutions, and universities provided the majority of personnel working on Apollo.

NASA also moved quickly during the early 1960s to expand its physical capabilities. In 1960 the space agency consisted of a small headquarters in Washington, D.C., three aeronautical research centers, the Jet Propulsion Laboratory, the Goddard Space Flight Center, and the Marshall Space Flight Center. With the advent of Apollo, these installations grew rapidly. In addition, NASA added three new facilities specifically to meet the demands of the lunar landing program. In 1962 it created the Manned Spacecraft Center (renamed the Lyndon B. Johnson Space Center in 1973) near Houston, Texas. This center also became the home of NASA's astronauts and the site of mission control. For the Apollo project, NASA greatly expanded the Launch Operations Center at Cape Canaveral on Florida's eastern seacoast. Renamed the John F. Kennedy Space Center on 29 November 1963, this installation's massive and expensive Launch Complex 39 was the site of all lunar launches. Additionally, the spaceport's Vehicle Assembly Building was a huge and expensive 36-story structure where the Saturn/Apollo rockets were assembled. Finally, to support the development of the Saturn launch vehicle, in October 1961 NASA created on a bayou in the deep South the Mississippi Test Facility, renamed the John C. Stennis Space Center in 1988.

HOW DO WE GO TO THE MOON?

One of the critical early decisions made by NASA was the method of going to the moon. Three basic approaches were advanced to accomplish the lunar mission:

1. *Direct Ascent* called for the construction of a huge booster that would launch a spacecraft, send it on a course directly to the moon, land a large vehicle, and send some part of it back to Earth. The Nova booster project, which was to have been capable of generating up to 40 million pounds of thrust, would have been able to accomplish this feat. Even if other factors had not impaired the possibility of direct ascent, the huge cost and technological sophistication of the Nova rocket quickly ruled out the option and resulted in cancellation of the project early in the 1960s despite the conceptual simplicity of the direct ascent

method. The method had few advocates when serious planning for Apollo began.

2. *Earth-Orbit Rendezvous* was the logical first alternative to the direct ascent approach. It called for the launching of various modules required for the moon trip into an orbit above Earth, where they would rendezvous and be assembled into a single system, refueled, and sent to the moon. This could be accomplished using the Saturn launch vehicle already under development by NASA that was capable of generating 7.5 million pounds of thrust. A logical component of this approach was the establishment of a space station in Earth orbit to serve as the lunar mission's rendezvous, assembly, and refueling point. In part because of this prospect, a space station emerged as part of the long-term planning of NASA as a jumping-off place for the exploration of space. This method of reaching the moon, however, was also fraught with challenges—notably, finding methods of maneuvering and carrying out a rendezvous in space, assembling components in a weightless environment, and safely refueling spacecraft.
3. *Lunar-Orbit Rendezvous* involved sending up the entire lunar spacecraft in one launch. It would head to the moon, enter into orbit, and dispatch a small lander to the lunar surface. It was the simplest of the three methods in terms of both development and operational costs, but it was risky. Since rendezvous would take place in lunar, instead of Earth, orbit, there was no room for error (otherwise the crew could not get home). Moreover, some of the trickiest course corrections and maneuvers had to be done after the spacecraft had been committed to a circumlunar flight. The Earth-orbit rendezvous approach kept open all the options for the mission longer than did the lunar-orbit rendezvous mode.

Inside NASA, advocates of the various approaches argued over the method of flying to the moon while the all-important clock that Kennedy had started continued to tick. It was critical that a decision not be delayed, because the mode of flight in part dictated the type of spacecraft developed. Although NASA engineers could proceed with building a launch vehicle (the Saturn) and define the basic components of the spacecraft—a habitable crew compartment, a baggage car of some type, and a jettisonable service module containing propulsion and other expendable systems—they could not proceed much beyond rudimentary conceptions without a mode decision. The NASA Rendezvous Panel, headed by John C. Houbolt, pressed hard for the lunar-orbit rendezvous as the most expeditious means of accomplishing the mission. Using sophisticated technical and economic arguments, over a period of months in 1961 and 1962 Houbolt's group persuaded the rest of NASA's leadership that lunar-orbit rendezvous was not the risky proposition it had seemed earlier.

The last to give in were Wernher von Braun and his associates at the Marshall Space Flight Center. This group favored the Earth-orbit rendezvous for three reasons: because the direct ascent approach was technologically unfeasible before the end of the 1960s, because it provided a logical rationale for a space station, and because it ensured an extension of the workload at the Marshall Center (a factor that was always important to center directors competing inside the agency for personnel and other resources). At an all-day meeting on 7 June 1962 at Marshall, NASA leaders met to hash out these differences, with the debate growing heated at times. After more than six hours of discussion, von Braun finally gave in to the lunar-orbit rendezvous model, saying that its advocates had demonstrated adequately its feasibility and that any further contention would jeopardize the president's timetable. This set the stage for the space flights to follow.

PRELUDE TO APOLLO: MERCURY

At the time of the announcement of Project Apollo by President Kennedy in May 1961, NASA was consumed with the task of placing an American in orbit through Project Mercury. Stubborn problems arose, however, at almost every turn. The first space flight of an astronaut, made by Alan B. Shepard, had been postponed for weeks so NASA engineers could resolve numerous details; it finally took place on 5 May 1961, less than three weeks before the Apollo announcement. The second flight, a suborbital mission like Shepard's, launched on 21 July 1961, also had problems. The hatch blew off prematurely from the Mercury capsule, *Liberty Bell 7*, and sank into the Atlantic Ocean before it could be recovered. In the process the astronaut, Virgil I. "Gus" Grissom, nearly drowned before being hoisted to safety in a helicopter. These suborbital flights, however, proved valuable for NASA technicians, who found ways to solve or work around literally thousands of obstacles to successful space flight.

As these issues were being resolved, NASA engineers began final preparations for the orbital aspects of Project Mercury. In this phase NASA planned to use a Mercury capsule capable of supporting a human in space not just for minutes, but eventually for as long as three days. As a launch vehicle for this Mercury capsule, NASA used the more powerful Atlas instead of the Redstone. But the decision was not without controversy. Indeed, there were technical difficulties to be overcome in mating it to the Mercury capsule, but the biggest complication was a debate among NASA engineers over its appropriateness for human space flight.

Most of the difficulties had been resolved by the time of the first successful orbital flight of an unoccupied Mercury-Atlas combination in September 1961. On 29 November the final test flight took place, this time with the chimpanzee Enos occupying the capsule for a two-orbit ride before being successfully recovered in an ocean landing. Not until 20 February 1962, however, could NASA accomplish an orbital flight with an astronaut. On that date John Glenn became the first American to circle the Earth, making three orbits in the *Friendship 7* Mercury spacecraft. The flight was not without problems, however; Glenn flew parts of the last two orbits manually because of an autopilot failure, and he left his retrorocket pack (which normally would be jettisoned) attached to the capsule during re-entry because of a loose heat shield.

Glenn's flight provided a healthy boost in national pride, making up for at least some of the earlier Soviet successes. The public, more than celebrating the technological success, embraced Glenn as a personification of heroism and dignity. Hundreds of requests for personal appearances by Glenn poured into NASA headquarters, and NASA learned much about the power of the astronauts to sway public opinion. The agency leadership allowed Glenn to speak at some events, but more often it substituted other astronauts and declined many other invitations. Among other engagements, Glenn did address a joint session of Congress and participated in several ticker-tape parades around the country. NASA thereby discovered a powerful public relations tool which it has employed ever since.

Three more successful Mercury flights took place during 1962 and 1963. Scott Carpenter made three orbits on 20 May 1962, and on 3 October 1962 Walter Schirra flew six orbits. The capstone of Project Mercury was the 15–16 May 1963 flight of Gordon Cooper, who circled Earth 22 times in 34 hours. The program succeeded in accomplishing its purpose: to orbit successfully a human in space, explore aspects of tracking and control, and learn about microgravity and other biomedical issues associated with space flight.

GEMINI: BRIDGING THE GAP BETWEEN MERCURY AND APOLLO

Even as the Mercury program was under way and work took place developing Apollo hardware, NASA program managers perceived a huge gap in the capability for human space flight between that acquired with Mercury and what would be required for a lunar landing. They closed most of the gap by experimenting and training on the ground, but some issues required experience in space. Three major issues were immediately evident. The first was the ability in space to locate, maneuver toward, and rendezvous and dock with another spacecraft. The second was closely related: the ability of astronauts to work outside a spacecraft. The third involved the collection of more sophisticated physiological data about the human response to extended space flight.

To gain experience in these areas before Apollo could be readied for flight, NASA devised Project Gemini. Initiated in the fall of 1961 by engineers at Robert Gilruth's Space Task Group in cooperation with McDonnell Aircraft Corporation technicians (builders of the Mercury spacecraft), Gemini started out as a larger Mercury *Mark II* capsule but soon became a totally different proposition. It could accommodate two astronauts for extended flights of more than two weeks. It pioneered the use of fuel cells instead of batteries to power the ship, and it incorporated a series of modifications to hardware. Its designers also considered using a paraglider being developed at Langley Research Center for "dry" landings instead of a "splashdown" in water and recovery by the Navy. The entire system was to be powered by the newly developed *Titan II* launch vehicle, another ballistic missile developed for the Air Force. A central reason for this program was to perfect techniques for rendezvous and docking, so NASA appropriated from the military some Agena rocket upper stages and fitted them with docking adapters.

Problems with the Gemini program abounded from the start. The *Titan II* had longitudinal oscillations, called the pogo effect because it resembled the behavior of a child on a pogo stick. Overcoming this problem required engineering imagination and long hours of overtime to stabilize fuel flow and maintain vehicle control. The fuel cells leaked and had to be redesigned, and the Agena reconfiguration also suffered costly delays. NASA engineers never did get the paraglider to work properly, and eventually they dropped it from the program in favor of a parachute system like the one used for Mercury. All these difficulties shot an estimated $350 million program to over $1 billion. The overruns were successfully justified by the space agency, however, as necessities to meet the Apollo landing commitment.

By the end of 1963 most of the difficulties with Gemini had been resolved and the program was ready for flight. Following two unoccupied orbital test flights, the first operational mission took place on 23 March 1965. Mercury astronaut Gus Grissom commanded the mission, with John W. Young, a naval aviator chosen as an astronaut in 1962, accompanying him. The next mission, flown in June 1965, stayed aloft for four days, and astronaut Edward H. White II performed the first American extravehicular activity (EVA), or spacewalk. Eight more missions followed through November 1966. Despite problems great and small encountered on virtually all of them, the program achieved its goals. Additionally, as a technological learning program, Gemini had been a success, with 52 different experiments performed on the 10 missions. The bank of data acquired from Gemini helped bridge the gap between Mercury and what would be required to complete Apollo within the time constraints directed by the president.

BUILDING SATURN

NASA inherited the effort to develop the Saturn family of boosters used to launch Apollo to the moon in 1960 when it acquired the Army Ballistic Missile Agency under Wernher von Braun. By that time, von Braun's engineers were hard at work on the first-generation Saturn launch vehicle, a cluster of eight Redstone boosters around a Jupiter fuel tank. This first rocket in the Saturn family was solely a research and development vehicle that would lead toward the accomplishment of Apollo.

The largest and most significant launch vehicle of this family, the *Saturn V*, represented the culmination of earlier booster development and test programs. Standing 363 feet tall, with three stages, this was the vehicle that took astronauts to the moon and returned them safely to Earth. The first stage generated 7.5 million pounds of thrust from five massive engines developed for the system. These engines, known as the F-1, were some of the most significant engineering accomplishments of the program, requiring the development of new alloys and different construction techniques to withstand the extreme heat and shock of firing. The thunderous sound of the first static test of this stage, which took place at Huntsville, Alabama, on 16 April 1965, brought home to many observers that the Kennedy goal was within technological grasp. For others, it signaled the magic of technological effort; one engineer even characterized rocket engine technology as a "black art" without rational principles. The second stage presented enormous challenges to NASA engineers and very nearly caused the lunar landing goal to be missed. Consisting of five engines burning liquid oxygen (LOX) and liquid hydrogen, this stage could deliver one million pounds of thrust. It was consistently behind schedule, and it required constant attention and additional funding to ensure completion by the deadline for a lunar landing. Both the first and third stages of this Saturn vehicle development program moved forward relatively smoothly.

To "prove" the technology, NASA adopted an "all-up" concept in which the entire Apollo-Saturn system was tested together in flight before astronauts flew it. A calculated gamble, the first *Saturn V* "all-up" test launch took place on 9 November 1967 with the entire Apollo-Saturn combination. A second test followed on 4 April 1968. Even though it was only partially successful because the second stage shut off prematurely and the third stage (needed to start the Apollo payload into lunar trajectory) failed, NASA declared that the test program had been completed and that the next launch would have astronauts aboard. The gamble paid off. In 17 test launches and 15 piloted launches, the Saturn booster family scored a 100 percent launch reliability rate.

THE APOLLO SPACECRAFT

With the announcement of the lunar landing commitment in 1961, NASA technicians began a crash program to develop a vehicle for the trip to lunar orbit and back. They came up with a three-person command module capable of sustaining human life for two weeks or more in either Earth orbit or a lunar trajectory; a service module holding oxygen, fuel, maneuvering rockets, fuel cells, and other expendable and life support equipment that could be jettisoned upon re-entry to Earth; a retrorocket package attached to the service module for slowing to prepare for re-entry; and a launch escape system that was to be discarded upon achieving orbit. The teardrop-shaped command module had two hatches: one on the side for entry and exit of the crew at the beginning and end of the flight, and one in the nose with a docking collar for use in moving to and from the lunar landing vehicle.

Work on the Apollo spacecraft stretched from 28 November 1961, when the prime contract for its development was let to North American Aviation, to July 1975, when the last Apollo spacecraft flight took place. In between there were various efforts to design, build, and test the spacecraft—both on the ground and in suborbital and orbital flights. By the end of 1966, NASA leaders declared the Apollo command module ready for human occupancy. The final flight checkout of the spacecraft prior to the lunar flight took place on 11–22 October 1968 with three astronauts.

BUILDING THE LUNAR MODULE

If the Saturn launch vehicle and the Apollo spacecraft were difficult technological challenges, the third part of the hardware for the moon landing, the lunar module (LM), represented the most complex problem. Begun a year later than originally scheduled, the LM was consistently behind schedule and over budget. Much of the problem turned on the demands of devising two separate spacecraft components—one for descent to the moon, and one for ascent back to the command module—that only maneuvered outside an atmosphere. Both engines had to work perfectly, or the very real possibility existed that the astronauts would not return home. Guidance, maneuverability, and spacecraft control also caused endless headaches. The landing structure likewise presented problems; it had to be light and sturdy and shock resistant. An ungainly vehicle emerged that two astronauts could fly while standing. Beginning in November 1962, Grumman Aerospace Corporation began work on the LM. With difficulty the LM was orbited on a *Saturn V* test launch in January 1968 and judged ready for operation.

THE TRAGEDY OF *APOLLO 204*

As these development activities were taking place, tragedy struck the Apollo program. On 27 January 1967, *Apollo-Saturn (AS) 204*, scheduled to be the first space flight with astronauts aboard the capsule, was on the launch pad at Kennedy Space Center, Florida, moving through simulation tests. The three astronauts who were to fly on this mission—"Gus" Grissom, Edward White, and Roger B. Chaffee—were aboard running through a mock launch sequence. At 6:31 P.M., after several hours of work, a fire broke out in the spacecraft and the pure oxygen atmosphere intended for the flight helped it burn with intensity. In a flash, flames engulfed the capsule and the astronauts died of asphyxiation. It took the ground crew five minutes to open the hatch. When they did so, they found three bodies. Although three other astronauts had been killed before this time—all in plane crashes—these were the first deaths directly attributable to the U.S. space program.

Shock gripped NASA and the nation during the days that followed. James Webb, NASA administrator, told the media at the time, "We've always known that something like this was going to happen soon or later. . . . Who would have thought that the first tragedy would be on the ground?"[4] The day after the fire, NASA appointed an eight-member investigation board that set out to discover the details of the tragedy: what happened, why it happened, whether it could happen again, what was at fault, and how NASA could recover.

The members of the board soon learned that the fire had been caused by a short circuit in the electrical system that ignited combustible materials in the spacecraft and that it had been fed by the pure oxygen atmosphere. They also found that it could have been prevented. Subsequently they called for several modifications to the spacecraft, including a move to a less oxygen-rich environment. Changes to the capsule followed thereafter, and within a little more than a year it was ready for flight.

NASA administrator James Webb reported these findings to various congressional committees and underwent a personal grilling at every meeting. His answers were sometimes evasive and always defensive. The *New York Times*, which was usually critical of Webb, had a field day with this situation and said that the NASA acronym stood for "Never a Straight Answer." Although the ordeal was personally taxing, whether by happenstance or design Webb deflected much of the backlash over the fire from both NASA as an agency and from the Johnson administration. Although he was personally tarred by the disaster, the space agency's image and popular support were largely undamaged. Webb himself never recovered from the stigma of the fire, and when he left NASA in October 1968, even as Apollo was nearing a successful completion, few mourned his departure.

Recovery from the *Apollo-Saturn 204* capsule fire took more than a year, but in October 1968 the Apollo system was flown by astronauts in Earth orbit. Again it looked like reaching the moon on Kennedy's timetable was a possibility. On 21 December 1968 *Apollo 8* took off atop a *Saturn V* booster from the Kennedy Space Center with three astronauts aboard—Frank Borman, James A. Lovell Jr., and William A. Anders—for a historic mission to orbit the moon. So far Apollo had been all promise; now the delivery was about to begin.

After *Apollo 8* made one-and-a-half Earth orbits, its third stage began, a burn to put the spacecraft on a lunar trajectory. As it traveled outward, the crew focused a portable television camera on Earth and for the first time humanity saw its home from afar, a tiny, lovely, and fragile "blue marble" hanging in the blackness of space. When *Apollo 8* arrived at the moon on Christmas Eve, this image of Earth was even more strongly reinforced when the crew sent back images of the planet while reading the first part of the Bible—"God created the heavens and the Earth, and the Earth was without form and void"—before sending Christmas greetings to humanity. The next day they fired the boosters for a return flight and "splashed down" in the Pacific Ocean on 27 December. It was an enormously significant accomplishment at a time when American society was in crisis over war in Vietnam, race relations, urban problems, and a host of other difficulties. If only for a few moments, the nation united as one to focus on this epochal event. Two more Apollo missions occurred before the climax of the program, but they did little more than confirm that the time had come for a lunar landing.

THE FLIGHT OF *APOLLO 11*

Then came the big event. *Apollo 11* lifted off on 16 July 1969 and, after confirming that the hardware was working well, began the three-day trip to the moon. At 4:18 P.M. EST on 20 July 1969, the LM—with astronauts Neil A. Armstrong and Edwin E. Aldrin—landed on the lunar surface while astronaut Michael Collins orbited overhead in the Apollo command module. After checkout, Armstrong set foot on the lunar surface, telling millions who saw and heard him on Earth that it was "one small step for [a] man—one giant leap for mankind."[5] (Neil Armstrong later added "a" when referring to "one small step for *a* man" to clarify the first sentence delivered from the moon's surface.) Aldrin soon followed him out, and the two plodded around the landing site in the lunar gravity (one-sixth that of Earth), planted an American flag (but omitted claiming the land for the United States, as had been routinely done during European exploration of the Americas), collected soil and rock samples, and set up scientific experiments. The next day they launched back to the Apollo capsule orbiting overhead and began the return trip to Earth, splashing down in the Pacific on 24 July 1969.

These flights rekindled the excitement felt in the early 1960s with John Glenn and the Mercury astronauts. *Apollo 11*, in particular, met with an ecstatic reaction around the globe, as everyone shared in the success of the mission. Ticker-tape parades, speaking engagements, public relations events, and a world tour by the astronauts served to create good will both in the United States and abroad.

COMPLETING APOLLO

Six more Apollo missions followed at approximately six-month intervals through December 1972, and with the exception of the aborted *Apollo 13* flight, each of them increased the time spent on the moon. Three of the later Apollo missions used a lunar rover vehicle to travel in the vicinity of the landing site, but none of them equaled the excitement of *Apollo 11*. The scientific experiments placed on the moon, and the lunar soil samples returned through Project Apollo, have provided grist for scientists' investigations of the solar system ever since. The scientific return was significant, but the Apollo program did not answer conclusively the age-old questions of lunar origins and evolution.

In spite of the success of the other missions, only *Apollo 13*, launched on 11 April 1970, came close to matching earlier public interest. But that was only because 56 hours into the flight, an oxygen tank in the Apollo service module ruptured and damaged several of the power, electrical, and life support systems. People throughout the world watched, waited, and hoped as NASA personnel on the ground and the crew—well on their way to the moon and with no way of returning until they went around it—worked together to find a way safely home. While NASA engineers quickly determined that air, water, and electricity did not exist in the Apollo capsule at sufficient levels to sustain the three astronauts until they could return to Earth, they found that the LM—a self-contained spacecraft unaffected by the accident—could be used as a "lifeboat" to provide austere life support for the return trip. It was a close call, but the crew returned safely on 17 April 1970. The near-disaster served several important purposes for the civil space program, especially prompting reconsideration of the appropriateness of the entire effort while also solidifying in the popular mind NASA's technological genius.

A MEANING FOR APOLLO

Project Apollo in general, and the flight of *Apollo 11* in particular, constituted a watershed in the nation's twentieth-century history. It was an endeavor that demonstrated both the technological and economic virtuosity of the United States and established its technological preeminence over rival nations—the primary goal of the program when first envisioned by the Kennedy administration in 1961.

Project Apollo left several important legacies (or conclusions). First, and probably most important, the Apollo program was successful in accomplishing the goals for which it had been created. At the time of the *Apollo 11* landing, Mission Control in Houston flashed on its big screen the words of President Kennedy announcing the Apollo commitment. Those phrases were followed with these: "TASK ACCOMPLISHED, July 1969." No greater understatement could have been made. Any assessment of Apollo that does not recognize the accomplishment of landing an American on the moon and safely returning before the end of the 1960s is incomplete and inaccurate, for that was the primary goal of the undertaking.

Second, Project Apollo was a triumph of management in meeting enormously difficult systems engineering, technological, and organizational integration requirements. James E. Webb, the NASA administrator at the height of the program between 1961 and 1968, always contended that Apollo was much more a management exercise than anything else, and that the technological challenge, while sophisticated and impressive, was largely within grasp at the time of the 1961 decision. More difficult was the task of ensuring that those technological skills were properly managed and used.

Webb's contention was more than confirmed by the success of Apollo. NASA leaders had to acquire and organize unprecedented resources to accomplish the task at hand. From both a political and technological perspective, management was critical. For seven years after Kennedy's Apollo decision, through October 1968, James Webb maneuvered for NASA in Washington to obtain sufficient resources to meet Apollo requirements. More to the point, NASA personnel employed the "program management" concept that emphasized centralized authority and systems engineering. The systems management of the program was critical to Apollo's success. Understanding the management of complex structures for the successful completion of a multifarious task was a critical outgrowth of the Apollo effort.

Third, Project Apollo forced the people of the world to view the planet Earth in a new way. *Apollo 8* was critical to this fundamental change, as it treated the world to the first pictures of Earth from afar. Writer Archibald MacLeish summed up the feelings of many people when he wrote at the time of Apollo that "To see the Earth as it truly is, small and blue and beautiful in that eternal silence where it floats, is to see ourselves as riders on the Earth together, brothers on that bright loveliness in the eternal cold—brothers who know now that they are truly brothers."[6] The modern environmental movement was galvanized in part by this new perception of the planet and the need to protect it and the life that it supports.

Finally, the Apollo program, despite being an enormous achievement, left a divided legacy for NASA and the aerospace community. The perceived "golden age" of Apollo created for the agency an expectation that the direction of any major space goal from the president would always bring NASA a broad consensus of support and provide it with the resources and license to dispense them as it saw fit. What most NASA officials did not understand at the time of the moon landing in 1969, however, was that Apollo had not been conducted under normal political circumstances and that the exceptional circumstances surrounding Apollo would not be repeated.

The Apollo effort was, therefore, an anomaly in the national decision-making process. The dilemma of the "golden age" of Apollo has been difficult to overcome, but moving beyond the Apollo program to embrace future opportunities has been an important goal of the agency's leadership at the end of the twentieth century. Exploration of the solar system and the universe remains as enticing a goal and as important an objective for humanity as it ever has been. Project Apollo was an important early step in that ongoing process of exploration.

CONCLUSION

While the United States engaged in a very public race to the moon in the 1960s, the Soviet Union's leadership claimed it was not trying to get there at all. Indeed, the Soviet leaders castigated American officials for heightening Cold War tensions with Apollo while they claimed peaceful intentions in a measured space exploration effort. But their public comments belied their secretive actions, because in fact Soviet rocket design bureaus and scientific institutions worked around the clock to beat the Americans to the moon. Senior Soviet rocket designers such as Sergei P. Korolev and Valentin P. Glushko designed hardware to travel to the moon and oversaw an increasingly sophisticated human space flight program that was planned eventually to entail the planting of the Soviet hammer and sickle on Earth's nearest neighbor.

Despite very significant Soviet successes in space during the decade, the fact that they failed to reach the moon can be attributed to two interrelated factors. First, the Soviet Union was not as technologically advanced as the United States and, despite illusions to the contrary in the West, did not have the necessary base to carry out successfully the most complex undertaking ever attempted by humankind. Nevertheless, their efforts were perhaps within 80–85 percent of attaining their objectives. The N-1 moon rocket, however, kept exploding—indicative of technical challenges present throughout the program. Second, the Soviet Union's space program never enjoyed the type of top-down management structure that NASA developed in the United States. This meant that program management, so critical to any large-scale technological endeavor, never achieved the primacy necessary to ensure full success. Individual space technology leaders warred among themselves and vied with Soviet leaders for favors of funding. The Soviet lunar program never had a James Webb to bring order, and the closest anyone could come to commanding that type of unity was Sergei Korolev. He died in January 1966, and the Soviet lunar program took a nosedive thereafter. Others tried to hold it together until the early 1970s; but eventually after American success and repeated Soviet failures, it was cancelled.

Even though American leaders knew from intelligence sources much of what the Soviet Union was doing in space, publicly the Soviet leaders denied any interest in reaching the moon. Only with the collapse of the Soviet Union and the end of the Cold War in the 1990s did the West learn the true extent of Soviet efforts. The American public then realized that the Soviet Union had lunar landers, spacecraft, and rockets constructed and waiting for use on a trip to the moon. But the Soviet capability remained only a prospect. The United States had realized its stated mission to land on the moon. In doing so, the United States not only reclaimed its pride in its technological and scientific "superiority" but, in the public mind, defined the terms of space exploration thereafter.

NOTES

1. Quoted in John M. Logsdon, *The Decision to Go to the Moon: Project Apollo and the National Interest* (Cambridge, MA: MIT Press, 1970), p. 111.
2. *New York Times*, 17 April 1961.
3. John F. Kennedy, "Urgent National Needs," Speech to a Joint Session of Congress, 25 May 1961, Presidential Files, Kennedy Library.
4. Quoted in Erik Bergaust, *Murder on Pad 34* (New York: G. P. Putnam's Sons, 1968), p. 23.
5. CBS News, *10:56:20 PM EDT, 7/20/69: The Historic Conquest of the Moon as Reported to the American People* (New York: Columbia Broadcasting System, 1970).
6. Quoted in Oran W. Nicks, ed., *This Island Earth* (Washington, DC: NASA SP-250, 1970), p. 3.

3

Scientific Explorations in the Space Age

From the 1960s to the present, to some Americans, space has represented prestige and the American image on the world stage. That outlook drove the U.S. effort to reach the moon before the Soviet Union did. To other Americans, space has signified the quest for national security. To still others, space is, or should be, a source of greater knowledge about the universe. In this outlook it represents pure science and exploration of the unknown. In the overall thinking and writing about space exploration, the first two views of American interest in space have eclipsed the third. Indeed, the history of space science itself is a largely neglected aspect in the history of the space program.

SPACE SCIENCE AND THE INTERNATIONAL GEOPHYSICAL YEAR

Although space science went back to the early twentieth century and advanced significantly during and after World War II (with the development of German V-2 rocketry, for example), the history of U.S. space science really began in 1957 when the first artificial satellites were launched as part of the International Geophysical Year (IGY). This effort began in 1955 when the U.S. National Security Council approved development of a small scientific satellite "under international auspices, such as the International Geophysical Year, in order to emphasize its peaceful purposes; . . . considerable prestige and psychological benefits will accrue to the nation which first is successful in launching a satellite . . . especially if the USSR were to be the first to establish a satellite."[1]

This commitment was followed by competition between the Naval Research Laboratory, and the Army's Redstone Arsenal for government support to develop an IGY satellite. Project Vanguard, proposed by the Navy, was chosen on 9 September 1955 to carry the standard in launching a nonmilitary satellite for the IGY effort; the Navy's project was chosen over the Army's "Explorer" proposal. The decision was made largely because the Naval Research Laboratory's candidate would not interfere with high-priority ballistic missile programs (it had the nonmilitary Viking rocket as its basis), whereas the Army's bid was heavily involved in those activities and proposed adapting a ballistic missile launch vehicle. In addition, the Navy project seemed to have greater promise for scientific research because of a larger payload capacity.

While Vanguard was in development, the Soviet Union launched *Sputnik I* on 4 October 1957 and utterly changed the nature of space science. This satellite was the Soviet entry into the IGY program, and its success spelled crisis in the United States. Within weeks accelerated efforts for American space flight had been placed in motion. To catch up, on 31 January 1958 the U.S. Army's Ballistic Missile Agency used a Redstone rocket to place the first American satellite, *Explorer 1*, into orbit. This satellite discovered what came to be known as the Van Allen belts, of radiation and a terrestrial magnetosphere.

Since *Explorer 1*, measurements by numerous spacecraft have calculated the size and intensity of this field, showing that it contains two belts of very energetic, charged particles. At its upper limits the magnetosphere encounters charged plasma particles thrown off by the sun, known collectively as the solar wind, which create a boundary where the two come into contact. These electromagnetic fields interact and create shock waves, some of which are responsible for auroral phenomena such as the aurora borealis, or northern lights. Perhaps as important as the discovery of the radiation belts and the rise of magnetospheric physics, *Explorer 1* demonstrated that a team of academic scientists could design and build instruments that worked in space.

A ROBOTIC RACE TO THE MOON

With the conclusion of the IGY, the United States went head-to-head with the Soviet Union in a robotic race to the moon—and lost. The moon was an early target for both the United States and the Soviet Union because it was comparatively close; in the context of cosmic distances it is the neighbor in the next apartment whose stereo can be heard through thin walls. There were also numerous opportunities every month for a launch from Earth to the moon, and it would be a significant public relations coup in the international community for the nation reaching it first.

The lure of the moon was irresistible. In 1958, eager to demonstrate leadership in space technology, the United States started a crash effort to send a series of spacecraft named Pioneer to the moon. The Air Force prepared three, and the Army built two Pioneer spacecraft for the moon flights. During the winter of 1958–1959 the United States made four attempts to send a Pioneer probe in the vicinity of the moon. None reached their destination. Indeed, none succeeded in escaping from Earth's orbit, but two produced the first information about the outer regions of the Van Allen belts.

In contrast, after some false starts in the fall of 1958, the Soviet Union succeeded in launching several successful probes. This success rested on the early capability of Soviet engineers to build large rockets with significant payload capacity, something not yet developed in the United States. In January 1959 the Soviets sent *Luna 1* past the moon and into orbit around the sun, following up with *Luna 3* to transmit pictures of the far side of the moon—thereby giving the Soviets an important "first" in lunar exploration. Meanwhile, in March 1959, *Pioneer 5* finally flew past the moon, much too late to assuage America's loss of pride and prestige. Thus ended the first phase of lunar exploration, with the Soviet Union a clear winner.

In December 1959, after the failure of the first lunar probes, the Jet Propulsion Laboratory (JPL) started the Ranger project, partly as a way to get out of the public relations mess the earlier failures had created. On 30 August 1961, NASA launched the first Ranger, but the launch vehicle placed it in the wrong orbit. Two more attempts in 1961 failed, as did two more attempts in 1962. NASA then reorganized the Ranger project and did not try to launch again until 1964. By this time its engineers had eliminated all the scientific instruments except a television camera. Ranger's sole remaining objective was to go out in a blaze of glory as it crashed into the moon while taking high-resolution pictures. Finally, on 31 July 1964, the seventh Ranger worked and transmitted 4,316 beautiful, high-resolution pictures of the lunar Sea of Clouds. The eighth and ninth Rangers also worked well.

Other projects by the United States followed, all designed to support Project Apollo's goals of landing an American on the moon before the end of the decade. Ultimately these lunar satellite exploration programs succeeded in providing data useful to the scientific community and information about the moon useful to those planning the Apollo landings, but only after significant missteps and false starts and considerable investment by the nation. They did, however, effectively pave the way for Apollo lunar landings, which also provided significant data about Earth's closest celestial neighbor. In so doing, the United States finally eclipsed the early successes of the Soviet Union in sending probes to the moon.

APOLLO LUNAR SCIENCE AND POST-APOLLO MISSIONS

Because of the commitment to accomplish Apollo as a Cold War goal, the human expedition to the moon was never conceived primarily as a science program, yet it yielded valuable scientific results almost as serendipity. Each mission carried an Apollo Lunar Surface Experiments Package (ALSEP). The experiments ranged from an effort to study the effects of general relativity on the motion of the moon, to a measurement of lunar dust. Most important, a geophysicist, Dr. Harrison H. (Jack) Schmitt, flew on the last Apollo mission and studied lunar geology in situ in what was the first (and thus far the only) fieldwork undertaken on the moon by a geologist. The lunar rock and soil samples returned by Apollo missions have been enormously significant. Those samples remain in pristine condition in a nitrogen environment at the Lunar Receiving Laboratory at the Johnson Space Center in Houston, Texas, available for study by scientists worldwide. In all, much excellent science was accomplished that had not been envisioned at the time of the political decision to carry out Apollo.

Throughout the 1960s and early 1970s when the United States was undertaking its crash efforts to explore the moon, the Soviet Union did not abandon its own efforts there. Indeed, although it failed to accomplish a human landing because of serious problems with its large booster development program exacerbated by management difficulties, the Soviet Union sent a large number of probes to the moon and returned significant scientific data about Earth's only natural satellite.[2]

The Soviet lunar program of the 1960s and early 1970s had 20 successful missions to the moon and achieved a number of notable lunar firsts: first probe to impact the moon, first flyby and image of the lunar farside, first soft landing, first lunar orbiter, and first circumlunar probe to return to Earth. The two successful series of Soviet probes were the Luna (15 missions) and the Zond (5 missions). Lunar flyby missions (*Luna 3; Zond 3, 6, 7*, and *8*) obtained photographs of the lunar surface, particularly the limb (southern) and farside regions. The *Zond 6, 7*, and *8* missions circled the moon and returned to Earth, where they were recovered (*Zond 6* and *7* in Siberia, and *Zond 8* in the Indian Ocean). The purpose of the photography experiments on the lunar landers (*Luna 9, 13*, and *22*) was to obtain close-up images of the surface of the moon for use in lunar studies and determination of the feasibility of piloted lunar landings. Three robotic missions (*Luna 16, 20*, and *24*) also soft-landed and returned lunar samples to Earth. Between the end of the Apollo program in December 1972 and the return of *Luna 24* in August 1976, the Soviets had the moon to themselves and flew three successful missions during this period.

The Soviet lunar sample return missions provided especially important scientific information. Scientists worldwide shared data and samples from both the Apollo and Luna missions, comparing results and geological details. For their part, the Soviets recognized that the Americans had been the first to return soil and rock samples to Earth but also stressed that their own efforts with automated probes had been more economical (the cost of the Soviet automated missions to the moon was only about 25 percent that of Apollo) and could land in more interesting geographical areas such as mountains because no humans were aboard. Nonetheless, Soviet space program analysts admitted that a dozen automated missions were required to equal the scientific return of a single Apollo landing mission.[3]

In 1994 the United States returned to the moon for the first time since Apollo. *Clementine* was a joint project between the Department of Defense's (DOD) Strategic Defense Initiative Organization and NASA. Its objective was to test sensors and spacecraft components under extended exposure to the space environment and to make scientific observations of the moon. The observations included imaging at various wavelengths, including ultraviolet and infrared, laser ranging altimetry, and charged particle measurements.

Operated on a shoestring budget, the *Clementine* spacecraft was launched on 25 January 1994 from the West Coast launch complex at Vandenburg Air Force Base, and it achieved lunar insertion on 21 February. Lunar mapping took approximately two months, but as the mission was beginning a new phase in May 1994, the spacecraft malfunctioned. In spite of this, *Clementine* mapped more than 90 percent of the lunar surface. The late 1996 revelation from scientists that data returned by *Clementine* indicated that ice existed from an asteroid crash at the moon's south pole re-energized lunar science, and additional missions are now under development that will explore this area further.

EXPLORATION OF THE SOLAR SYSTEM

During the 1960s, both the United States and the Soviet Union began an impressive effort to gather information about the planets of the solar system by using ground-, air-, and space-based equipment. Especially important was the creation of two types of spacecraft: one a probe that was sent toward a heavenly body, and the second an Earth-orbiting observatory that could gain the clearest resolution available in telescopes because it did not have to contend with the atmosphere. Once again, the compilation of this new data revolutionized humanity's understanding of Earth's immediate planetary neighbors. Although the most significant findings of this investigation would not come until the 1970s (perhaps the "golden age" of planetary science), studies of the planets captured the imagination of people from all backgrounds like nothing else except the Apollo lunar missions. For all the genuine importance of magnetospheric physics and solar studies, it was photographs of the planets and theories about the origins of the solar system that appealed to a broad cross-section of the public. As a result, NASA had little difficulty in capturing and holding a widespread interest in this aspect of the space science program.[4]

EXPLORATION OF VENUS

Planetary exploration began, just as lunar exploration had, in a race between the United States and the Soviet Union to see who would be the first to place some sort of spacecraft near Venus. This was not just an opportunity to best the rival in the Cold War; scientists in both the United States and the Soviet Union recognized the attraction of Venus for the furtherance of planetary studies as well. Regarded as both the evening and the morning star, Venus had long enchanted humans—and all the more so since astronomers had realized that it was shrouded in a mysterious cloak of clouds permanently hiding the surface from view. It was also the closest planet to Earth, and a near twin to this planet in terms of size, mass, and gravitation.

These attributes brought myriad speculations about the nature of Venus and the possibility of life existing there in some form. For instance, in the first half of the twentieth century a popular theory held that the sun had gradually been cooling for millennia and that as it did so, each planet in the solar system had a turn as a haven for life of various types. Although it was now Earth's turn to harbor life, the theory suggested that Mars had once been habitable and that life on Venus was now just beginning to evolve. Beneath the clouds of the planet, the theory offered, was a warm, watery world and the possibility of aquatic and amphibious life. "It was reasoned that if the oceans of Venus still exist, then the Venusian clouds may be composed of water droplets," noted JPL researchers; "if Venus were covered by water, it was suggested that it might be inhabited by Venusian equivalents of Earth's Cambrian period of 500 million years ago, and the same steamy atmosphere could be a possibility."[5]

After carrying out ground-based efforts in 1961 to view the planet using radar, which could "see" through the clouds, and learning among other things that Venus rotated in a retrograde motion opposite from the direction of orbital motion, both the Soviet Union and the United States began a race to the planet with robotic spacecraft. The Soviets tried first, launching *Venera 1* on 12 February 1961. Unlike lunar exploration, however, the Soviets did not win the race to Venus; their spacecraft broke down on the way. The United States claimed the first success in planetary exploration during the summer of 1962 when *Mariner 1* and *Mariner 2* were launched toward Venus. Although *Mariner 1* was lost during a launch failure, *Mariner 2* flew by Venus on 14 December 1962 at a distance of 21,641 miles. It probed the clouds, estimated planetary temperatures, measured the charged particle environment, and looked for a magnetic field similar to Earth's magnetosphere (but found none). Most important, it found that the planet's surface was a fairly uniform 800 degrees Fahrenheit, thereby making unlikely the theory that life—at least as humans understood it—existed on Venus.

Although the Soviet Union made several more attempts to reach Venus, only in 1965 was it successful in reaching the surface when *Venera 3* crashed there without returning any scientific data. In 1967 the Soviets sent *Venera 4*, which successfully deployed a probe into the atmosphere and returned further information about the makeup of the planet's surface. In the same year the Americans sent *Mariner 5* to Venus to investigate its atmosphere. Both spacecraft demonstrated that Venus was a very inhospitable place for life to exist. Collectively, these and subsequent planetary probes revealed that Venus was superheated because of the greenhouse effect of the cloud layer and that the pressure on the surface was about 90 atmospheres, far greater than even in the depths of the oceans on Earth.

Because of Venus's thick cloud cover, scientists early on advocated sending a probe with radar to map Venus. *Pioneer 12* had made a start toward realizing this goal, orbiting the planet for more than a decade to complete a low-resolution radar topographic map. Likewise, the Soviets' *Venera 15* and *16* missions in 1983 provided high-resolution coverage over the northern reaches of the planet. But the best opportunity to learn the features of the Venusian surface came in the early 1990s with what turned out to be a highly successful *Magellan* mission to Venus. Launched in 1989, *Magellan* mapped 99 percent of the surface at high resolution, parts of it in stereo. This data provided some surprises; among them the discoveries that plate tectonics was at work on Venus and that lava flows showed clear evidence of volcanic activity. Although years of research await Venus explorers from the data returned by these probes, collectively they fundamentally altered most of the beliefs held as recently as a generation ago about Venus as a tropical, proto-organic planet.

THE LURE OF THE RED PLANET

Like Venus, Mars had long held a special fascination for humans who pondered the planets of the solar system—partly because of the possibility that life might either presently exist or at some time in the past have existed there. Percival Lowell became interested in Mars during the latter part of the nineteenth century, and he built what became the Lowell Observatory near Flagstaff, Arizona, to study the planet. He argued that Mars had once been a watery planet and that the topographical features known as canals had been built by intelligent beings. The idea of intelligent life on Mars remained in the popular imagination for a long time, and only with the scientific data returned from probes to the planet since the beginning of the space age did this begin to change.

In June 1963 the Soviets reached Mars first, but with little scientific return. The United States did not reach Mars until 15 July 1965, when *Mariner 4* flew within 6,118 miles of the planet and took 21 close-up pictures. These photographs dashed the hopes of many that life might be present on Mars, for the first close-up images showed a cratered, lunar-like surface. They depicted a planet without structures and canals, nothing that even remotely resembled a pattern that intelligent life might produce.

Mariner 6 and *Mariner 7*, launched in February and March 1969, each passed Mars in August 1969, studying its atmosphere and surface to lay the groundwork for an eventual landing on the planet. Their pictures verified the moon-like appearance of Mars and gave no hint that Mars had ever been able to support life.

There was still hope, however, and the search for signs of life prompted emphasis on the exploration of Mars. NASA administrator James C. Fletcher, for example, commented on this possibility in 1975:

> Although the discoveries we shall make on our neighboring worlds will revolutionize our knowledge of the Universe, and probably transform human society, it is unlikely that we will find intelligent life on the other planets of our Sun. Yet, it is likely we would find it among the stars of the galaxy, and that is reason enough to initiate the quest.... We should begin to listen to other civilizations in the galaxy. It must be full of voices, calling from star to star in a myriad of tongues. Though we are separate from this cosmic conversation by light years, we can certainly listen ten million times further than we can travel....
>
> It is hard to imagine anything more important than making contact with another intelligent race. It could be the most significant achievement of this millennium, perhaps the key to our survival as a species.[6]

Not everything was rosy with Mars exploration, however. Despite delays, Project Viking represented the culmination of a series of exploratory missions that had begun in 1964 with *Mariner 4* and continued with *Mariner 6* and *Mariner 7* flybys in 1969 and a *Mariner 9* orbital mission in 1971 and 1972. The Viking mission used two identical spacecraft, each consisting of a lander and an orbiter. Launched on 20 August 1975 from the Kennedy Space Center in Florida, *Viking 1* spent nearly a year cruising to Mars, placed an orbiter in operation around the planet, and landed on 20 July 1976 on the Chryse Planitia (Golden Plains). *Viking 2* was launched on 9 September 1975 and landed on 3 September 1976. The

Viking project's primary mission ended on 15 November 1976, 11 days before Mars's superior conjunction (its passage behind the sun), although the Viking spacecraft continued to operate for six years after first reaching Mars. The last transmission from the planet reached Earth on 11 November 1982.

One of the most important scientific activities of this project involved an attempt to determine whether there was life on Mars. Although the three biology experiments discovered unexpected and enigmatic chemical activity in the Martian soil, they provided no clear evidence for the presence of living microorganisms in soil near the landing sites. According to mission biologists, Mars was self-sterilizing. They concluded that the combination of solar ultraviolet radiation that saturates the surface, the extreme dryness of the soil, and the oxidizing nature of the soil chemistry had prevented the formation of living organisms in the Martian soil. However, the question of life on Mars at some time in the distant past remains open.

In 1988 the Soviet Union, just a year away from collapse and the end of the Cold War, sent *Phobos 1* and *2* to Mars, but both failed enroute. The Americans' *Mars Observer* fared little better. Intended as a mapping mission, NASA lost contact with *Mars Observer* on 21 August 1993, three days before the spacecraft's capture in orbit around Mars. The loss of the *Mars Observer*, which represented an investment of nearly $1 billion, probably occurred as a result of an explosion in the propulsion system's tanks as they were pressurized.

Mars exploration received additional impetus in August 1996, however, when a team of NASA and Stanford University scientists announced that a Mars meteorite found in Antarctica contained possible evidence of ancient Martian life. When the 4.2-pound, potato-sized rock (identified as ALH84001) was formed as an igneous rock about 4.5 billion years ago, Mars was much warmer and probably contained oceans hospitable to life. Then, about 15 million years ago, a large asteroid hit the red planet and jettisoned the rock into space, where it remained until it crashed into Antarctica around 11,000 B.C. Scientists presented three compelling, but not conclusive, pieces of evidence suggesting that fossil-like remains of Martian microorganisms, which date back 3.6 billion years, are present in ALH84001. The findings electrified the scientific world but excited the public just as fully, and added support for an aggressive set of missions to Mars by the year 2000 to help discover the truth of these theories.[7]

The exploration received new life on 4 July 1997, when *Mars Pathfinder* successfully landed on Mars, the first return to the red planet since 1976. Its small, 23-pound robotic rover, named *Sojourner*, departed the main lander and began to record weather patterns, atmospheric opacity, and the chemical composition of rocks washed down into the Ares Vallis flood plain, an ancient outflow channel in Mars' northern hemisphere. This vehicle completed its projected milestone 30-day mission on 3 August 1997, capturing far more data on the atmosphere, weather, and geology of Mars than scientists had expected. In all, the *Pathfinder* mission returned more than 1.2 gigabits (1.2 billion bits) of data and over 10,000 tantalizing pictures of the Martian landscape.

A new portrait of the Martian environment began to emerge in the months since *Pathfinder* because of the new data that was collected in this mission. The rover's alpha proton X-ray spectrometer team analyzed the first-ever in-situ measurements of Mars rocks. Similarly, atmospheric-surface interactions, measured by a meteorology package onboard the lander, are confirming some conditions observed by the *Viking* landers twenty-one years ago, while raising questions about other aspects of the planet's global system of transporting volatiles such as water vapor, clouds, and dust. The meteorology mast on the lander observed a rapid drop-off in temperature just a few feet above the surface, and one detailed 24-hour measurement set revealed temperature fluctuations of 30–40 degrees Fahrenheit in a matter of minutes.

Moreover, *Sojourner*, a robust rover capable of semi-autonomous "behaviors," captured the imagination of the public, which followed the mission with great interest via the World Wide Web. Twenty *Pathfinder* mirror sites recorded 565 million hits worldwide during the period of 1 July to 4 August 1997. The highest volume of hits in one day occurred on 8 July, when a record 47 million hits were logged, which is more than twice the volume of hits received on any one day during the 1996 Olympic Games in Atlanta. The data from the mission is still being analyzed and will help to answer some of the age old questions associated with the origins of the solar system.

A GRAND TOUR OF THE OUTER SOLAR SYSTEM

As the space flight projects of the 1960s—culminating in the lunar exploration effort—gave way to more constrained budgets in the 1970s, NASA's leadership conceived their most daring planetary science expedition ever. Once every 176 years, the giant planets on the outer reaches of the solar system all gather on one side of the sun, and such a configuration was due to occur in the late 1970s. This geometric lineup made possible close-up observation of all the planets in the outer solar system (with the exception of Pluto) in a single flight, the so-called Grand Tour. The flyby of each planet would bend the spacecraft's flight path and increase its velocity enough to deliver it to the next destination. This would occur through a complicated process known as "gravity assist," something like a slingshot effect, whereby the flight time to Neptune could be reduced from 30 to 12 years.

To prepare the way for the Grand Tour, in 1964 NASA conceived *Pioneer 10* and *11* as outer solar system probes. Although severe budgetary constraints prevented starting the project until the fall of 1968 and forced a somewhat less ambitious effort, *Pioneer 10* was launched on 3 March 1972. It arrived at Jupiter on the night of 3 December 1973, and although many were concerned that the spacecraft might be damaged by intense radiation discovered in Jupiter's orbital plane, the spacecraft survived, transmitted data about the planet, and continued on its way out of the solar system, away from the center of the Milky Way galaxy. By May 1991 it had reached about 52 Astronautical Units (AU), roughly twice the distance from Jupiter to the sun, and was still transmitting data. It was anticipated that it will be sending back data through the end of the twentieth century. The next time it is intended to make contact again is in 1999.

In 1973 NASA launched *Pioneer 11*, which would provide scientists with their closest view of Jupiter. The close approach and the spacecraft's speed of 107,373 mph, by far the fastest ever reached by an object launched from Earth, hurled *Pioneer 11* 1.5 billion miles across the solar system toward Saturn, encountering the planet's south pole within 26,600 miles of its cloud tops in December 1974. In 1979 *Pioneer 11* again encountered Saturn, this time closing to within 13,000 miles of the planet, where it discovered two new moonlets and a new ring, and charted the magnetosphere, its magnetic field, its climate and temperatures, and the general structure of Saturn's interior. In 1990 *Pioneer 11* officially departed the solar system by passing beyond Pluto and headed into interstellar space toward the center of the Milky Way galaxy. Both *Pioneer 10* and *11* were remarkable space probes, stretching from a 30-month design life cycle into a mission of more than 20 years and returning useful data not just about the jovian planets of the solar system but also about some of the mysteries of the interstellar universe.[8]

Meanwhile, NASA technicians prepared to launch what became known as Voyager. Even though the four-planet mission was known to be possible, it soon became too expensive to build a spacecraft that could go the distance, carry the instruments needed, and last long enough to accomplish such an extended mission. Thus, the two Voyager spacecraft were funded to conduct intensive flyby studies only of Jupiter and Saturn, in effect repeating on a more elaborate scale the flights of the two Pioneers. Nonetheless, the engineers designed as much longevity into the two Voyagers as the $865 million budget would allow. NASA launched them from the Kennedy Space Center, Florida: *Voyager 2* lifted off on 20 August 1977, and *Voyager 1* entered space on a faster, shorter trajectory on 5 September 1977.

As the mission progressed, having successfully accomplished all its objectives at Jupiter and Saturn in December 1980, additional flybys of the two outermost giant planets, Uranus and Neptune, proved possible—and irresistible—to mission scientists. Accordingly, as the two spacecraft flew across the solar system, remote-control reprogramming was used to redirect the Voyagers for the greater mission. Eventually *Voyager 1* and *Voyager 2* explored all the giant outer planets, 48 of their moons, and the unique systems of rings and magnetic fields those planets possess.

The two spacecraft returned information to Earth that has revolutionized solar system science, helping resolve some key questions while raising intriguing new ones about the origin and evolution of the planets. The two Voyagers took well over 100,000 images of the outer planets, rings, and satellites, as well as millions of magnetic, chemical spectra, and radiation measurements. They discovered rings around Jupiter, volcanoes on Io, shepherding satellites in Saturn's rings, new moons around Uranus and Neptune, and geysers on Triton. The last imaging sequence was *Voyager 1*'s portrait of most of the solar system, showing Earth and six other planets as sparks in a dark sky lit by a single bright star, the sun.

RETURN TO JUPITER

It was nearly two decades after Voyager before any spacecraft ventured to the outer solar system again. In October 1989, NASA's *Galileo* spacecraft began a gravity-assisted journey to Jupiter, where it sent a probe into the atmosphere and observed the planet and its satellites for two years beginning in December 1995. Jupiter was of great interest to scientists because it appeared to contain material in an original state left over from the formation of the solar system, and the mission was designed to investigate the chemical composition and physical state of Jupiter's atmosphere and satellites. Because of a unique orbital inclination that sent the probe around the sun and back, on the way to Jupiter *Galileo* came back past both Venus and Earth and made the first close flyby of asteroid Gaspra in 1991, providing scientific data on all. But the mission was star-crossed. Soon after deployment from the space shuttle, NASA engineers learned that *Galileo*'s umbrella-like, high-gain antenna could not be fully deployed. Without this antenna, communication with the spacecraft was more difficult and time-consuming, and data transmission was greatly hampered. The engineering team working on the project tried a series of cooling exercises designed to shrink the antenna central tower and enable its deployment. Over a period of several months the engineers worked on this maneuver repeatedly but were unable to free the antenna. Through the end of 1995 the spacecraft's performance and condition were excellent except that the high-gain antenna was still only partly deployed; science and engineering data were therefore being transmitted via the much slower and less effective low-gain antenna, which the mission team had decided to use for the Jupiter encounter.

In mid-1995 *Galileo* deployed the probe that would parachute into Jupiter's dense atmosphere. The two spacecraft then flew in formation the rest of the way to Jupiter; while the probe began its descent into the planet's atmosphere, the main spacecraft went into a trajectory that placed it in a near-circular orbit. On 7 December 1995 the probe began its descent. Its instruments began relaying back data to the orbiter on the chemical composition of the atmosphere, the nature of the cloud particles and structure of the cloud layers, the atmosphere's radiative heat balance and pressure and dynamics, and the ionosphere. The probe lasted for about 45 minutes before the atmosphere and the pressure of the planet destroyed it. During that time the orbiter stored the returned data. With the high-gain antenna being inoperative, for months thereafter scientists and technicians coaxed the data back to Earth for analysis. Until the late 1990s *Galileo* will continue to transmit scientific measurements back to Earth for analysis. The result promises a reinterpretation of human understanding about Jupiter and its moons.

INVESTIGATION OF THE UNIVERSE

At the same time that these findings were reshaping knowledge of the solar system, space scientists were investigating and also profoundly affecting humanity's understanding of the universe beyond. The traditional field of astronomy underwent a tremendous burst of activity with the advent of the space age because of the ability to study the stars through new types of telescopes. In addition to greatly enhanced capabilities for observation in the visible light spectrum, NASA and other institutions have supported the development of a wide range of X-ray, gamma ray, ultraviolet, infrared, microwave, cosmic ray, radar, and radio astronomical projects. These efforts have collectively informed the most systematic efforts yet to explain the origins and development of the universe.

Before the space age, all astronomy was performed from the ground and was limited by the Earth's atmosphere, which filtered out many types of energy and some visible light sources. The stars and galaxies could only be seen in the visible light and radio spectrums, and large telescopes were constructed to observe at these wavelengths. Using these instruments, in 1927 Edwin P. Hubble, an astronomer at the Mount Wilson Observatory near Pasadena, California, discovered that other galaxies were apparently receding from the Milky Way, our own galaxy, and that the further away they were, the faster they retreated. Other astronomers built on Hubble's discovery to create a theory that the universe had originated at what has been called the Big Bang and it has been expanding at a constant rate for 10 to 20 billion years. Alternative theories to the Big Bang also have been advanced, and much of the exploration of the universe since Hubble has been oriented toward acquiring information that might conclusively prove one or another of these theories. The debate has continued to the present, and it is one of the truly exciting aspects of space science, involving fundamental questions of life and meaning.

The space age provided an opportunity to expand far beyond the capabilities offered by the observatories of Hubble's time. Fundamental to this was the development of a series of orbiting observatories, first conceptualized not long after the establishment of NASA as a means of creating an economy of scale with large spacecraft. Instead of using a variety of small satellites and several classes of launch vehicles for space science, these satellites consolidated many kinds of experiments on large observatories launched by Atlas-class rockets. In the 1960s, these observatories were only marginally successful. Each required an assemblage of booms, rotating platforms, and special apertures to satisfy the scientific requirements. Scientists and engineers spent much time and effort to eliminate interference among the various experiments and between the experiments and the spacecraft.

These efforts have been ongoing since the beginnings of the space age and represent key developments in the expansion of human knowledge about the universe. By the early 1970s, satellite astronomy had helped generate a major change in the larger field of astronomy and had reordered thinking on the subject. This occurred in spite of the fact that much of the research was built on the foundations laid by Edwin Hubble and other earlier astronomers. The interplay of energy and matter on a cosmological level has been enormously exciting for many Americans, as research findings have been reported in the public media. Moreover, although a wide variety of scientific fields enjoyed the yield of research data obtained from the new tools available to scientists, during the 1970s two important disciplines began to emerge as foremost in the field: the exploration of the solar system, and the study of the universe. Throughout this era funding for space science and applications in NASA was never more than $760 million per year (and usually much less).

THE GREAT SPACE TELESCOPES

In the mid-1960s, NASA's Office of Space Science started a second major space science project, the Apollo Telescope Mount (ATM), as a follow-on to the lunar landing mission and to capitalize on the hardware being developed for it. Initially NASA planned to start work on an unmanned Advanced Orbiting Solar Observatory (AOSO); however, budget limitations forced postponement. The human space flight program was looking for payloads for what was to become *Skylab*, the first space station. Ample funds existed in the manned space flight budget for experiments that could use *Skylab*. Taking advantage of this opportunity, AOSO became ATM, the Apollo Telescope Mount.

NASA launched *Skylab* and the ATM on 14 May 1973. ATM was much more than a telescope mount. It was a full-size solar observatory with sophisticated instruments, continuously pointed at the sun. Scientist-astronauts on board *Skylab* selected the targets on the sun and controlled the instruments. Some instruments used film, thereby improving spatial and spectral resolution. The crew monitored active regions of the sun; when they saw signs of flare they focused the instruments on the region, turned on the cameras, and obtained a continuous record of the radiation emitted during the flare. In addition to recording solar flares in action, the astronauts repaired several instruments during the flight, enabling all the instruments to work throughout the six-month experimental life of *Skylab*.

In 1965 the Space Science Board recommended that NASA begin work on a large, diffraction-limited space telescope that would take full advantage of the Saturn launch vehicle. Thus, NASA started research on the pointing system for the large system that eventually became the Hubble Space Telescope (HST). Initially NASA planned to build a four-meter mirror, the largest mirror that would fit inside the shroud of a Saturn upper stage. Later, to take advantage of existing national security optical production facilities, NASA reduced the size of the mirror to 3.2 meters. With the beginning of space shuttle flights in the 1980s, NASA redesigned the telescope so that it could be launched by the shuttle and serviced in space. Ultimately the $2 billion HST was launched from the space shuttle in April 1990, and astronomers were excited that it represented a quantum leap forward in astronomical capability. Through the telescope they expected to see with much greater resolution than ever before, viewing galaxies as far away as 15 billion light years. A key component of the telescope was a precision-ground, 94-inch primary mirror shaped to within microinches of perfection from ultra-low-expansion titanium silicate glass with an aluminum-magnesium fluoride coating.

The Hubble Space Telescope had been scheduled for launch in 1986 but was delayed during the space shuttle redesign that followed the *Challenger* accident in January of that year. Excitement abounded as it was finally deployed four years later and the first images began to come back to Earth. The photos provided bright, crisp images against the black background of space, much clearer than pictures of the same target taken by ground-based telescopes. Controllers then began moving the telescope's mirrors to better focus the images. Although the focus sharpened slightly, the best image still had a pinpoint of light encircled by a hazy ring, or halo. NASA technicians concluded that the telescope had a "spherical aberration," a mirror defect only one/twenty-fifth the width of a human hair, that prevented the HST from focusing all light to a single point.

At first many believed that the spherical aberration would cripple the

43-foot-long telescope, and NASA received considerable negative publicity. But soon scientists found a way with computer enhancement to work around the abnormality, and engineers planned a shuttle repair mission to correct it with an additional instrument. By 1993 the Hubble Space Telescope was returning impressive scientific data on a routine basis. For instance, as recently as 1980 astronomers had believed that an astronomical grouping known as R-136 was a single star, but the HST showed that it was made up of more than 60 of the youngest and heaviest stars ever viewed. The dense cluster, located within the Large Magellanic Cloud, was about 160,000 light years from Earth, roughly 5.9 trillion miles away.

Because of the difficulties with the mirror of the Hubble Space Telescope, in December 1993 NASA launched the shuttle *Endeavour* on a repair mission to insert corrective equipment into the telescope and to service other instruments. During a week-long mission, *Endeavour*'s astronauts conducted a record five spacewalks and successfully completed all programmed repairs to the telescope. The first reports from the newly repaired HST indicated that the images being returned now were more than an order of magnitude (10 times) greater than those obtained before. The result has been enormously important to scientific understanding of the cosmos.

Because of the servicing mission, the HST dominated space science activities throughout the next year. The results from Hubble touched on some of the most fundamental astronomical questions of the twentieth century, including the existence of black holes and the age of the universe. Highlights of the Hubble Space Telescope results included:

> Compelling evidence for a massive black hole in the center of a giant elliptical galaxy located 50 million light years away. This observation provided very strong support for predictions made 80 years ago in Albert Einstein's general theory of relativity. Observations of great pancake-shaped disks of dust, raw material for planet formation, swirling around at least half of the stars in the Orion Nebula, the strongest proof yet that the process which may form planets is common in the universe. Confirmation of a critical prediction of the Big Bang theory, that the chemical element helium should be widespread in the early universe. The detection of this helium by HST may mark the discovery of a tenuous plasma that fills the vast volumes of space between the galaxies, the long-sought intergalactic medium. In October 1994, astronomers announced measurements that showed the universe to be between 8 and 12 billion years old, far younger than previous estimates of up to 20 billion years. These measurements were the first step in a three-year systematic program to measure accurately the scale, size and age of the universe.[9]

These discoveries continued in 1995 and 1996. For instance, scientists using the HST obtained the clearest images yet of galaxies that formed when the universe was a fraction of its current age. These pictures provided the first clues to the historical development of galaxies and suggested that elliptical galaxies developed remarkably rapidly into their present shapes.

However, spiral galaxies that existed in large clusters evolved over a much longer period—the majority were built and then torn apart by dynamic processes in a restless universe. The HST also discovered a new dark spot on Neptune, imaged the Eagle nebula in search of information about star formation, and observed the spectacular crash of Comet Shoemaker-Levy 9 into the planet Jupiter in 1994.

PROJECTIONS OF SPACE SCIENCE IN THE TWENTY-FIRST CENTURY

As space science activities edged toward the end of the twentieth century, many people were excited by prospects for the future. At the same time, debates about accomplishments and priorities, as well as costs, remained unresolved. However, these concerns pointed up the difficulty of building a constituency for large science and technology programs in a democracy. The rocky course of American and other spacefaring nations' projects provides important lessons about the nature of high-technology public policy. These projects were striking examples of what social scientists have called heterogeneous engineering, a concept that recognizes that scientific and technological issues are simultaneously organizational, social, economic, and political. Various interests often clash in the decision-making process as difficult calculations have to be made. These interests could potentially come together to make it possible to develop a project that would satisfy the majority of the priorities served by the political process, but at the same time many other interests would undoubtedly be left unsatisfied.

NOTES

1. National Security Council, NSC 5520, "Draft Statement of Policy on U.S. Scientific Satellite Program," 20 May 1955, copy in NASA Historical Reference Collection, History Office, NASA Headquarters, Washington, DC.

2. John M. Logsdon and Alain Dupas, "Was the Race to the Moon Real?" *Scientific American* 270 (June 1994): 36–43.

3. Boris Petrov, "The Case for Space Automation," *Pravda*, 24 September 1970; "Back to the Moon," *Time*, 29 January 1973.

4. Indeed, Representative George E. Brown Jr. (D-California) remarked in a speech at the National Academy of Sciences in Washington on 12 February 1992, that space science held the imagination of the public, even when the NASA budget was declining and spectacular missions were not being undertaken. "It is also important to recall that some of our proudest achievements in the space program have been accomplished within a stagnant, no growth budget," he noted. "The development of the Landsat program, the Viking lander, Voyagers I and II, Pioneer-Venus, and even the Space Shuttle were all carried out during the 1970s

when the NASA budget was flat. It would be wise to review how we set priorities and managed programs during this productive time."

5. Jet Propulsion Laboratory, *Mariner: Mission to Venus* (New York: McGraw-Hill, 1963), p. 5. This became an enormously popular conception in science fiction literature. See the 1949 short story by Arthur H. Clark, "History Lesson," in *Expedition to Earth* (New York: Ballantine Books, 1953), pp. 73–82, for an explanation of the theory.

6. James C. Fletcher, *NASA and the "Now" Syndrome* (Washington, DC: National Aeronautics and Space Administration, 1975), p. 7.

7. David H. Onkst, "Life on Mars and Europa? NASA Reveals Possible Evidence of Extraterrestrial Existence," *Space Times: Magazine of the American Astronautical Society* 35 (September–October 1996): 4–7.

8. Richard O. Fimmel, James A. Van Allen, and Eric Burgess, *Pioneer: First to Jupiter, Saturn, and Beyond* (Washington, DC: NASA SP-446, 1980); Richard O. Fimmel, William Swindell, and Eric Burgess, *Pioneer Odyssey* (Washington, DC: NASA SP-396, 1977); NASA Press Release, "Pioneer 11 to End Operations after Epic Career," 29 September 1995, NASA Historical Reference Collection.

9. "Hubble Space Telescope Scientific Results in 1994," *Space Times: Magazine of the American Astronautical Society* 34 (March–April 1995): 11.

4

A Permanent Presence in Space

In the late 1960s many in the leadership of the American space program realized that the abundant resources that had been made available for Project Apollo would not be available again. They turned instead to advocating the development of major projects that would create for the United States a permanent presence in space and, eventually, the capability to leave Earth permanently. This involved the development of an orbital workshop leading to a space station and a reusable vehicle to transport people and cargo to and from Earth-orbit in a reasonably efficient way. These became the Skylab and space station programs, as well as the space shuttle, which fundamentally affected the course of space exploration through the end of the twentieth century.

THE DESIRE FOR A SPACE STATION

From virtually the beginning of the twentieth century, those interested in the human exploration of space have viewed as central to that endeavor the building of a massive, Earth-orbital space station that would serve as the jumping-off point to the moon and the planets. Space exploration enthusiasts have always believed that a permanently occupied space station was a necessary outpost in the new frontier of space. The more technically minded recognized that once humans had achieved Earth-orbit about 250 miles up (the presumed location of any space station), the problem of gravity would be conquered and humans would be about halfway to anywhere else they might want to go in space.

Numerous theorists had studied the possibility of establishing a space station in Earth-orbit even before the beginning of the twentieth century. They promoted a vision of human destiny to explore the solar system and the central role of a space station in facilitating this goal. Accordingly, studies of space station configurations had been an important part of NASA planning in the 1960s. NASA scientists and engineers pressed for these studies because a space station would meet the needs of the agency for an orbital laboratory, observatory, industrial plant, launching platform, and drydock. However, the station was forced to the bottom of the priority heap in 1961 following President Kennedy's decision to land an American on the moon by the end of the decade. Once that mandate had been issued, there was no time to develop a space station in spite of the fact that virtually everyone involved in space exploration recognized its necessity. It resurfaced as a major follow-on effort even while Apollo was under way in the 1960s, but it took several unusual turns before emerging as the principal project of the agency.

SKYLAB: A PRELIMINARY SPACE STATION

Without clear presidential leadership in the late 1960s, NASA began to forge ahead on its own with whatever plans it could get approved for a continuation of U.S. space flight in the 1970s. One of these used Apollo technology to realize (at least partially) the longstanding dream of a space station. What NASA built was a relatively small orbital space platform, called *Skylab*, that could be tended by astronauts. NASA officials hoped it would be the precursor of a real space station.

The Skylab program originated in the 1960s to prove that humans could live and work in space for extended periods, and to expand knowledge of solar astronomy beyond what could be achieved from Earth-based observations. It made extensive use of Saturn and Apollo equipment by incorporating a reconfigured and habitable third stage of the *Saturn V* rocket as the basic component of the orbital station. NASA engineers developed two concepts for this space platform. The first was dubbed the "wet" approach, in which a Saturn upper stage that had been used to achieve orbit would be refurbished for habitation by astronauts in space. This concept proved too risky and difficult, so NASA opted for a "dry" concept in which the stage was completely outfitted as an orbital workshop before launch. Carried out on a remarkably small budget for a human spaceflight program (made possible in large measure because of the use of equipment developed and built with Project Apollo funding), the direct Skylab expenditure was less than $3 billion.

The 100-ton *Skylab 1* orbital workshop was launched into orbit on 14 May 1973, making use of the giant *Saturn V* launch vehicle for the last time. Almost immediately, technical problems developed due to vibrations during lift-off. Sixty-three seconds after launch, the meteoroid shield—designed also to shade *Skylab*'s workshop from the sun's rays—ripped off, taking with it one of the spacecraft's two solar panels, and another piece wrapped around the other panel to keep it from properly deploying. In spite of this, the space station achieved a near-circular orbit at the desired altitude of 270 miles. NASA's mission control personnel maneuvered *Skylab* so that its Apollo Telescope Mount (ATM) solar panels faced the sun to provide as much electricity as possible; but because of the loss of the meteoroid shield, this positioning caused workshop temperatures to rise to 126 degrees Fahrenheit.

While NASA technicians worked on a solution to the problem, *Skylab 2*, the first mission with astronauts aboard, was postponed. In an intensive ten-day period, NASA developed procedures and trained the crew to make the workshop habitable. At the same time, engineers "rolled" *Skylab* to lower the temperature of the workshop. Finally, on 25 May 1973 astronauts Charles Conrad Jr., Paul J. Weitz, and Joseph P. Kerwin lifted off from Kennedy Space Center in an Apollo capsule atop a Saturn IB and rendezvoused with the orbital workshop. This crew carried a parasol, tools, and replacement film to repair the orbital workshop. After substantial repairs requiring extravehicular activity (EVA), including deployment of a parasol sunshade that cooled the inside temperature to 75 degrees Fahrenheit, by 4 June the workshop was habitable. During a 7 June EVA, the crew also freed the jammed solar array and increased power to the workshop.

While in orbit the crew conducted solar astronomy and Earth resources experiments, medical studies, and five student experiments. This first crew made 404 orbits and carried out experiments for 392 hours, in the process making three EVAs totaling 6 hours and 20 minutes. The first group of astronauts returned to Earth on 22 June 1973, and two other Skylab missions followed. The first of these, *Skylab 3*, was launched using Apollo hardware on 28 July 1973 and its mission lasted 59 days. *Skylab 4*, the last mission on the workshop, was launched on 16 November 1973 and remained in orbit for 84 days. At the conclusion of *Skylab 4* the orbital workshop was powered down with the intention that it would be visited again in four years.

During the three missions, a total of three Apollo crews had occupied the orbital workshop for a total of 171 days and 13 hours. In Skylab, both the total hours in space and the total hours spent in performance of EVA under microgravity conditions exceeded the combined totals of all the world's previous space flights up to that time. NASA was also delighted with the scientific knowledge gained about long-duration space flight during the Skylab program, despite the workshop's early and recurring mechanical difficulties. It was the site of nearly 300 scientific and technical experiments.

Following the final occupied phase of the Skylab mission, ground controllers performed some engineering tests of certain Skylab systems (tests that ground personnel were reluctant to do while astronauts were aboard), positioned the orbital workshop into a stable attitude, and shut down its systems. It was expected that *Skylab* would remain in orbit for eight to ten years, by which time NASA might be able to reactivate it. In the fall of 1977, however, agency officials determined that *Skylab* had entered a rapidly decaying orbit—resulting from greater-than-predicted solar activity—and that it would re-enter the Earth's atmosphere within two years. They steered the orbital workshop so that debris from re-entry would fall over oceans and unpopulated areas of the planet. On 11 July 1979, *Skylab* finally hit the Earth. The debris scattered from the southeastern Indian Ocean across a sparsely populated section of western Australia. NASA and the U.S. space program took criticism for this development, ranging from the sale of hardhats as "Skylab Survival Kits" to serious questions about the propriety of space flight altogether if people were likely to be killed or injured by falling space debris. It was an inauspicious ending to the first American space station, but the experiment had whetted the appetite of NASA leaders for a permanent presence in space.

THE SPACE SHUTTLE DECISION

Because a space station would be too expensive, in the early 1970s NASA went on to build the space shuttle. This vehicle had originally been intended merely as a logistical craft to travel between Earth and a space station, but now it would become a substitute. The economics of space flight were critical to understanding the shuttle decision, because of the nature of the Nixon administration and the situation in the United States at the time. George M. Low, NASA's deputy administrator, said in a memorandum to the NASA leadership on 27 January 1970, "I think there is really only one objective for the space shuttle program, and that is 'to provide a low-cost, economical space transportation system.' To meet this objective, one has to concentrate both on low development costs and on low operational costs."[1] From the outset, therefore, the economics of the shuttle outweighed any other aspects of the program. This was a striking difference from the Apollo programs.

When first envisioned, the shuttle was to have been a two-stage, fully reusable system with both stages piloted and capable of landing on a runway like conventional aircraft. Launched like a rocket, the two stages would separate near the edge of Earth's atmosphere, with the first stage (about the size of a Boeing 747) returning to Earth. The second stage (about the size of a Boeing 707) would fly on under its own power into orbit, perform its mission, and then return to Earth. NASA reduced the cost of this system to an estimated $10 billion, but this was still too high for the president's approval. The agency went back to the drawing board and in 1971 come forward with a design that could be built for the projected bargain price of $5.5 billion.

On 5 January 1972, President Richard M. Nixon announced the decision to build a space shuttle. It came as a relief both to the aerospace industry and to space advocates. Both praised it as a great step forward in national capability. Critics derided it as an ill-timed, ill-considered, unnecessary expenditure of public funds. Supporters—especially Caspar W. Weinberger, who later became Nixon's defense secretary—argued that building the shuttle would reaffirm America's superpower status and help restore confidence, at home and abroad, in America's technological genius and will to succeed.

The space shuttle that emerged in the early 1970s consisted of three primary elements: a delta-winged orbiter spacecraft with a large crew compartment, a cargo bay 15 by 60 feet in size, and three main engines; two solid rocket boosters (SRBs); and an external fuel tank housing the liquid hydrogen and oxidizer burned in the main engines. The orbiter and the two solid rocket boosters were reusable. The shuttle was designed to transport approximately 45,000 tons of cargo into near-Earth orbit, 115 to 250 statute miles above the Earth. It could also accommodate a flight crew of up to ten persons (although a crew of seven would be more common) for a basic space mission of seven days. During a return to Earth, the orbiter was designed so that it had a cross-range maneuvering capability of 1,265 statute miles to meet requirements for lift-off and landing at the same location after only one orbit.

There were several challenges in building this entirely new type of space vehicle. Perhaps the most important design issue, after the orbiter's configuration, concerned the boosters to be developed and whether they should burn liquid or solid fuel. Also important was the development of the unique re-entry method of the shuttle orbiter. A question arose over how best to pass through the ionosphere: with a high angle of attack that would bring the orbiter through it quickly and heat the outer skin to extremely high temperatures (but only for a short period of time), or by using a blunt-body approach like that of earlier capsules. NASA eventually decided on an approach that required development of a special ceramic tile to be placed on the underside and nose of the orbiter to withstand the re-entry heat. Because of these issues, as well as political and management questions, shuttle development slowed down considerably in the mid-1970s, prompting its redefinition and refinancing and a delay of its first operational flight from 1978 to 1981. Meanwhile, *Skylab* collapsed in a fiery streak across the Pacific, and the United States was back to square one with its efforts to build and maintain a space station.

FIRST FLIGHT OF THE SHUTTLE *COLUMBIA*

With much public excitement, *Columbia*, the first orbiter that could be flown in space, took off from Kennedy Space Center, Florida, on 12 April 1981, six years after the last American astronaut had returned from orbit following a cooperative U.S./USSR Apollo-Soyuz test project in 1975. This first shuttle flight was piloted by astronauts John W. Young and Robert L. Crippen, one a veteran in space flight and the other a key member of the shuttle flight test team that had conducted atmospheric tests on the *Enterprise*.

At launch, the orbiter's three liquid-fueled engines—drawing propellants from the external tank—and the two SRBs generated approximately 7 million pounds of thrust. After about two minutes, at an altitude of 31 miles, the two boosters were spent and separated from the external tank. Waiting ships recovered them for eventual refurbishment and reuse on later missions. The spacecraft's main engines continued to fire for about eight minutes more before shutting down just as the shuttle entered orbit. As they did so, the external tank separated from the orbiter and followed a ballistic trajectory back to the ocean but was not recovered.

The orbiter reached a velocity on orbit of approximately 17,322 statute miles per hour, circling the globe in less than two hours. Once in orbit, Young and Crippen tested the spacecraft's on-board systems, fired the Orbital Maneuvering System (OMS) used for changing orbits and the Reaction Control System (RCS) engines used for attitude control, and opened and closed the payload bay doors (the bay was empty for this first test mission).

After 36 orbits during two days in space, excitement permeated the nation once again as *Columbia* landed like an aircraft at Edwards Air Force Base in California, successfully accomplishing the first such landing of an orbital vehicle in the history of the space age. The first flight had been an enormous success, and with it the United States had embarked on a new era of human space flight. Many considered the flight of *Columbia* as the beginning of an age in which there would be inexpensive and routine access to space for many people and payloads. Speculation abounded that within a few years shuttle flights would take off and land as predictably as airplanes, and that commercial tickets would be sold for regularly scheduled "spaceline" flights.

SHUTTLE OPERATIONS: A NEW DAWN?

High hopes abounded for the space shuttle following that first launch of *Columbia* in 1981, most of them over finally achieving the goal of low-cost, routine access to space. During the 24 flights that followed through mid-January 1986, one successful and spectacular launch after another took place. Although the primary mission of any spacecraft designed to carry astronauts must be to ensure the safety of its crew (not the undertaking of scientific experiments or the achieving of some other technical objective), the space shuttle was used as a platform for a number of important scientific endeavors.

Shuttle advocates often cited these activities as important contributions that justified the cost of the program. They especially endorsed the development of *Spacelab*, a sophisticated laboratory built by the European Space Agency that fit into the shuttle's cargo bay. Nonetheless, many scientists questioned the viability of the shuttle for scientific activities and suggested that the developmental costs could more usefully have been applied to expendable systems and robotic probes that promised higher scientific returns on investments.

Without question, the shuttle greatly expanded the opportunity for human space flight. By January 1996, the number of individuals flown stood at 181 men and 26 women. Among other notable developments, in June 1983 Dr. Sally K. Ride, a NASA scientist-astronaut, became the first American woman to fly in space aboard *STS-7*, and in August 1983 Guion S. Bluford became the first African-American astronaut to fly in space by serving on the crew of *STS-8*. The shuttle era also saw flights by people who were not truly astronauts. Senator Jake Garn (R-Utah) and Representative Bill Nelson (D-Florida) both left Congress long enough to fly on the shuttle in 1985 and 1986, respectively. After his flight, Nelson, chair of the House space science and applications subcommittee, offered this assessment of the space program: "If America ever abandoned her space ventures, then we would die as a nation, becoming second-rate in our own eyes, as well as in the eyes of the world. . . . Our prime reason for commitment can be summed up as follows . . . space is our next frontier."[2] In addition, the first teacher to fly in space would have been Christa McAuliffe, who died in the *Challenger* accident in January 1986; and astronauts from other nations flew aboard the shuttle during other missions. Even though this expansion of access to space flight was positively received in many quarters, some critics accused NASA of pandering to Congress and other constituencies for support by offering such perquisites to a carefully selected few.

In spite of the high hopes that had attended the first launch of *Columbia* in 1981, the shuttle program ultimately provided neither inexpensive nor routine access to space. By January 1986 there had been only 24 shuttle flights, although in the 1970s NASA had projected more flights than that for every year. Although the system was reusable, its complexity, coupled with the ever-present rigors of flying in an aerospace environment, meant that the turnaround time between flights was several months instead of several days. In addition, missions were delayed for a wide range of problems associated with ensuring the safety and performance of such a complex system. Since the flight schedule did not meet expectations, and since it took thousands of work hours and expensive parts to keep the system performing satisfactorily, observers began to criticize NASA for failing to meet the cost-effectiveness expectations that had been highlighted to gain federal approval for the shuttle program ten years earlier.

Critical analyses agreed that the shuttle had proven to be neither cheap nor reliable, both original selling points, and that NASA should never have used those arguments in building a political consensus for the program. Therefore, in some respects there was some agreement by 1985 that the effort had been both a triumph and a tragedy. The program had been engagingly ambitious and had developed an exceptionally sophisticated vehicle, one that no other nation could have built at the time. As such, it had been an enormously successful program. At the same time the shuttle was essentially a continuation of space spectaculars, as Apollo had been, and its much-touted capabilities had not been realized. It made far fewer flights and conducted far fewer scientific experiments than NASA had publicly predicted.

THE *CHALLENGER* ACCIDENT

Add to this the tragic loss of the space shuttle *Challenger* during launch from the Kennedy Space Center on 28 January 1986, and the difficulties of the shuttle program could not be ignored. Many Americans had been excited about this mission, even more than those that had gone before, because a member of the crew was a teacher who would be conducting a class from orbit. The Teacher in Space program, with the young and energetic Christa McAuliffe as its centerpiece, had been years in the making and was touted as a major step forward in education for young people. But the mission ended abruptly and tragically 73 seconds into the flight: a leak in one of two solid rocket boosters ignited the main liquid fuel tank, and *Challenger* exploded in a blazing fireball.

The accident, the worst in the history of the American space program, proved all the more devastating not only because of McAuliffe's and her crewmembers' deaths but also because of the close connections that other members of the crew had to groups in the United States. The seven crewmembers of the *Challenger* represented a richly diverse cross-section of the American population in terms of race, gender, geography, background, and religion. The explosion became one of the most significant events of the 1980s, as millions around the world saw the accident on television and mourned the crewmembers killed.

Following the *Challenger* accident, several investigations took place to understand the causes of the tragedy and to ascertain what changes should be made to the program to ensure shuttle safety and reliability. The most important investigation was the presidentially mandated blue ribbon commission chaired by former secretary of state William P. Rogers. The commission grappled with the technologically difficult issues associated with the *Challenger* accident, firmly linking it to a poor engineering decision made years earlier to use O-rings to seal joints in the SRBs. These, they found, were susceptible to failure at low temperatures. The Rogers Commission also criticized the communication system inside NASA, finding that concerns about irregularities in the O-rings had been voiced well before 1986, but that because of poor internal communication these concerns had not been raised to the appropriate level. In the words of one Rogers Commission member, NASA was "playing Russian roulette" because as long as the shuttle returned safely, the irregularities did not seem to affect successful operations.[3] Only in hindsight did the O-ring problem appear so daunting that it required the cancellation of a launch.

On 6 June 1986, the Rogers Commission submitted its formal report to President Ronald Reagan. The report included nine recommendations for restructuring the shuttle program and safely returning to flight. Most important, the SRBs were extensively redesigned following the accident. This involved recertifying the boosters through a series of test firings at Morton Thiokol's SRB facility in Brigham City, Utah. The redesign added an extra O-ring to the joints between the booster segments and greatly strengthened the physical connection between the segments. Heaters were added to the joints to prevent low temperatures from affecting the sealing capability of the O-rings.

In addition, NASA made extensive landing safety improvements. This included upgrades of the orbiter fleet's tires, brakes, and nose-wheel steering mechanism; and adding a drag chute system. NASA engineers also made other safety improvements, including the installation of a crew escape system that allowed astronauts to parachute from the orbiter under certain conditions. Moreover, the space shuttle program was completely reorganized to ensure that all necessary information would be available to managers at all levels, including a means of raising problems anonymously from any level of the program staff. Also in line with the Rogers Commission findings, experienced astronauts were placed in senior management positions within the program. Finally, through a series of open reviews, all significant issues were to be elevated to a Flight Readiness Review Board chaired by the NASA Associate Administrators for Space Flight and Safety. Through this process, NASA leaders hoped to foster full and open discussions of potential safety and operational issues.

A RETURN TO FLIGHT

When the space shuttle resumed flight operations with the launch of *Discovery* on 29 September 1988, it was a much safer program than it had been before the January 1986 accident. Through December 1997 there have been 61 shuttle missions since the *Challenger* accident. Counting all shuttle missions, including the *Challenger*, there have been 87 flights. Once flight began again in 1988, the vehicle returned to its former status as a workhorse of space exploration. Since 1988 the shuttle has launched the *Magellan* spacecraft to Venus, the *Galileo* spacecraft to Jupiter, and the *Ulysses* spacecraft to study the sun. The shuttle also has deployed the Gamma Ray Observatory, the Hubble Space Telescope, and the Upper Atmosphere Research Satellite. Between April 1981 and January 1997 the shuttle has carried approximately 2.3 million pounds of cargo and more than 700 major payloads into orbit for commercial interests, other nations, and educational institutions. Its crews have also conducted more than 45 extravehicular activities (spacewalks).

At the end of the twentieth century, the space shuttle enjoys the same praises and suffers from the same criticisms that have been voiced since shortly after the program first began. It remains the only vehicle in the world with the dual capability to deliver and return large payloads to and from orbit. The design, now more than two decades old, is still state-of-the-art in many respects, including computerized flight control, airframe design, electrical power systems, thermal protection system, and main engines. It is also the most reliable launch system now in service anywhere in the world, with a success rate of more than 98 percent.

At the same time, it is extremely expensive to fly and has been unable to deliver on its promise of routine access to space. Costing between $400 million and $1 billion for every flight, the shuttle program should be replaced by a modern launch system that will be more economical. Indeed, the inability of the space shuttle to meet the nation's space launch needs was emphasized in 1990 in a report by a presidentially appointed Advisory Committee on the Future of U.S. Space Programs, headed by aerospace corporation Martin-Marietta chief executive officer Norman R. Augustine. The report stated that "the most significant deficiency in the nation's future civil space program is an insufficiency of reliable, flexible, and efficient space launch capability."[4] Spaceflight visionaries were correct to emphasize low-cost, routine access to space. Unfortunately, NASA failed to achieve the goal with the space shuttle, and as the new century approaches, it is apparent that NASA must begin to develop a new vehicle to achieve that end. Without it, access to space will be severely limited indefinitely. With it, the planets might become accessible. Yet even though the problem has been diagnosed, there is little consensus about how to resolve it as the century closes.

THE SOVIET UNION'S SPACE STATION EFFORT

Like the Americans, Soviet space advocates had pressed for the development of a space station for that nation's space program almost from the beginning of the space age. As early as 1962, Soviet engineers proposed a space station composed of modules launched separately and brought together in orbit. Even as the United States was still completing Project Apollo, in 1971 the Soviet Union launched its first space station. Its first-generation space stations had one docking port and could not be resupplied or refueled. There were two types: highly secret Almaz military stations, and a publicly known set of Salyut civilian stations.

The Almaz military station program was the first to be approved. When it was proposed in 1964, it had three parts: the Almaz military surveillance space station, transport logistics spacecraft for delivering soldier-cosmonauts and cargo, and Proton rockets for launching both. All these spacecraft were built, but none was used as originally planned. To counter American success with Apollo, Soviet leaders directed that Almaz hardware be transferred to the civilian Salyut program so that the Soviet Union could recover a measure of international prestige with a spectacular public success. *Salyut 1*, the first space station in history, reached orbit atop a Proton rocket on 19 April 1971, two years before NASA's *Skylab*.

Those early first-generation stations were plagued by failures. The crew of *Soyuz 10*, the first spacecraft sent to *Salyut 1*, was unable to enter the space station because of a docking mechanism problem. The *Soyuz 11* crew lived aboard *Salyut 1* for three weeks but died during return to Earth because the air escaped from their *Soyuz* spacecraft. Then, three first-generation stations failed to reach orbit or broke up in orbit before crews could reach them. The Soviets recovered rapidly from these failures, though. *Salyut 3, Salyut 4*, and *Salyut 5* supported a total of five crews. In addition to military surveillance and scientific and industrial experiments, the cosmonauts performed engineering tests to help develop the second-generation space stations.

With the second-generation stations, the Soviet space station program evolved from short-duration to long-duration stays. These stations had two docking ports to permit refueling and resupplying of spacecraft. A second docking port also meant that long-duration resident crews could receive visitors. Visiting crews often included cosmonaut-researchers from Soviet-bloc countries or countries sympathetic to the Soviet Union. Vladimir Remek of Czechoslovakia, the first space traveler from neither the United States nor the Soviet Union, visited *Salyut 6* in 1978.

In the latter 1970s the Soviet Union expanded its capability by launching more capable stations. The first of these was *Salyut 6*, which orbited between 1977 and 1982. This station received a total of 16 cosmonaut crews, including 6 long-duration crews, and hosted cosmonauts from Hungary, Poland, Romania, Cuba, Mongolia, Vietnam, and East Germany.

The Soviet Union launched *Salyut 7* in 1982. Although it remained in orbit until 1991, its last crew flew in 1986. A near twin of *Salyut 6*, this station was home to 10 cosmonaut crews, including 6 long-duration crews, one of which set a record of orbiting for 237 days. The Soviet Union expanded its crew complement significantly during the operational life of *Salyut 7*, inviting France and India to send astronauts and flying the first Russian woman since 1963. *Salyut 7* was abandoned in 1986—when the Soviet Union began operating its first long-duration space station, *Mir*—and re-entered Earth's atmosphere over Argentina in 1991.

TO BUILD A REAL SPACE STATION

With *Skylab* gone from the scene after 1979, and the coming on line of the space shuttle as a system in 1981, NASA returned to its quest for a real space station as a site of orbital research and a jumping-off point to the planets during the early 1980s. In a measure of political skill not often seen previously, NASA administrator James M. Beggs persuaded President Ronald Reagan, against the wishes of many of his advisors, to endorse the building of a permanently occupied space station. In a "Kennedyesque" moment in 1984, Reagan declared that "America has always been greatest when we dared to be great. We can reach for greatness again. We can follow our dreams to distant stars, living and working in space for peaceful, economic, and scientific gain. Tonight I am directing NASA to develop a permanently manned space station and to do it within a decade."[5]

The Reagan administration certainly viewed the space station as a part of its larger strategy to defeat the Soviet Union, the "evil empire" in Reagan's words. That strategy involved a buildup of U.S. military capabilities; a confrontational approach to foreign policy; assistance to U.S. allies around the world, such as aid to rebels in Afghanistan; and the development of a multifaceted set of new technological systems. Those ranged from the building of a 600-ship navy and "stealth" aircraft that could evade enemy radar, to the Strategic Defense Initiative (SDI) involving the deployment of sophisticated space-based systems that could defeat Soviet missiles launched against the United States.

In this context the Reagan administration's objectives for Space Station *Freedom*, as it was called, required (1) an advancement in American technological know-how so that spin-off systems might also be used in SDI, and (2) the need for the space station to serve as a rallying point for the nation's allies. As a result, from the outset the space station had to be a high-priority international program. Although a range of international cooperative activities had been carried out in the past (*Spacelab*, the Apollo-Soyuz Test Project, and scientific data exchange), the station offered an opportunity for a truly integrated effort. It was thought that the inclusion of international partners—many of whom by now had their own rapidly developing spaceflight capabilities—could enhance the effort. In addition, every partnership would bring greater legitimacy to the overall program and might help insulate it from drastic budgetary and political changes. U.S. diplomats and politicians would not want to incite an international incident because of a change to the space station, and that fact might help stabilize funding, schedules, or other factors that might otherwise be changed in response to short-term political needs.

NASA leaders understood these issues but recognized that having international partners would also dilute their own authority to execute the program as they saw fit. Previously, the space agency had not been required to deal with partners, either domestic or international, as co-equals. It tended to see them more as a hindrance than a help, especially when they might block the "critical path" toward any technological goal. If international partners might be assigned responsibility for the development of a critical subsystem, NASA would be giving up the authority to make changes, to dictate solutions, and to control schedules and other factors. Partnership, furthermore, was not a synonym for contractor management (something the agency leaders understood very well), and NASA was not very accepting of full partners unless they were essentially silent or deferential. Such an attitude prevented NASA from wanting to take part in significant cooperation in large-scale programs.

In addition, some technologists expressed fear that bringing Europeans into the project would mean giving foreign nations technical knowledge that only the United States held. No other nation could build a space station on a par with *Freedom*, and only a handful had a genuine launch capability. Thus, many government officials questioned the advisability of reducing America's technological lead. The control of technology transfer in the international arena was an especially important issue to be considered.

Nonetheless, NASA leaders pressed forward with international agreements among 13 nations to take part in the Space Station *Freedom* program. In 1985 Japan, Canada, and nations pooling their resources in the European Space Agency (ESA) agreed to participate. Canada, for instance, decided to build a remote servicing system. Building on its *Spacelab* experience, ESA agreed to build an attached, pressurized science module and an astronaut-tended free-flyer. Japan's contribution was the development and commercial use of an experimental module for materials processing, life sciences, and technological development. These separate components, with their "plug-in" capacity, eased somewhat the overall management (and congressional) concerns about unwanted technology transfer.

Almost from the outset, the Space Station *Freedom* program was controversial. Most of the debate centered on its costs versus its benefits. One NASA official remembered that "I reached the scream level at about $9 billion," referring to how much U.S. politicians appeared willing to spend on the station.[6] As a result, NASA designed the project to fit an $8 billion research and development funding profile. For many reasons (some of them associated with Washington politics), within five years the projected costs had more than tripled and the station had become too expensive to fund fully—especially considering that the national debt had exploded in the 1980s.

NASA tried to reduce the space station budget, in the process eliminating functions that some of its constituencies wanted. This led to a rebellion among certain former supporters. For instance, the space science community complained that the space station configuration under development did not provide sufficient experimental opportunity. Thomas M. Donahue, an atmospheric scientist from the University of Michigan and chair of the National Academy of Sciences' Space Science Board, commented in the mid-1980s that his group "sees no scientific need for this space station during the next twenty years." He also suggested that "if the decision to build a space station is political and social, we have no problem with that," alluding to the thousands of jobs associated with it. "But don't call it a scientific program."[7]

Redesigns of Space Station *Freedom* followed in 1990, 1991, 1992, and 1993. Each time the project got smaller, less capable of accomplishing the broad projects originally envisioned for it, less costly, and more controversial. As costs were reduced, capabilities also were diminished, and increasingly political leaders who had once supported the program questioned its viability. It was a seemingly endless circle, and political pundits wondered when the dog would wise up and stop chasing its tail. Some leaders suggested that the nation, NASA, and the overall space exploration effort would be better off if the space station program were terminated.

Yet Congress did not terminate the program, in part because of the desperate economic situation in the aerospace industry (the result of an overall recession and of military demobilization after the collapse of the Soviet Union and the end of the Cold War) and in part because of the fact that by 1992 the project had spawned an estimated 75,000 jobs in 39 states. Politicians were hesitant to cancel the station outright because of these jobs, but neither were they willing to fund it at the level required to make it a truly viable program. As Barbara Mikulski (D-Maryland), chair of the Senate appropriations subcommittee that handled NASA's budget, said, "I truly believe that in space station *Freedom* we are going to generate jobs today and jobs tomorrow—jobs today in terms of the actual manufacturing of space station *Freedom*, but jobs tomorrow because of what we will learn."[8]

In the late 1980s and early 1990s, a parade of space station managers and NASA administrators—each of them sincere in their attempts to rescue the program—wrestled with *Freedom*. They faced, on one side, politicians demanding that the jobs aspect of the effort be maintained; on the other side, potential station users demanding that *Freedom*'s capabilities be maintained; and, on all sides, people demanding that costs be reduced. The incompatibility of these demands made station program management an extraordinarily difficult task. The NASA administrator who took office on 1 April 1992, Daniel S. Goldin, was faced with a uniquely frustrating situation when these competing claims were made official by the new president, Bill Clinton. In the spring of 1993, Clinton advised Goldin to restructure the space station program by reducing its budget, maximizing its scientific use, and ensuring that losses in aerospace industry jobs would be minimal.

After months of work, NASA came forward with three redesign options for the space station. On 17 June 1993, President Clinton decided to proceed with a moderately priced, moderately capable station design. On 7 November 1993, because of a dramatically changed international situation, the United States invited Russia to join in the building of the international space station. Nonetheless, the 14-nation international space station program remains a difficult issue as the 1990s progress and public policymakers wrestle without consensus over competing political agendas.

A centerpiece of the newly restructured space station effort was *Mir*, the Soviet Union's space station that has been in orbit since 1986. Weighing 20.4 tons at launch, by the early 1990s the complex weighed more than 70 tons because of additional modules; it now consists of the *Mir* core and the *Kvant, Kvant 2*, and *Kristall* modules. *Mir* is more than 107 feet long with docked *Progress-M* and *Soyuz-TM* spacecraft, and about 90 feet wide across its modules. Although the *Mir* core resembles *Salyut 7*, it has six ports—with those fore and aft used primarily for docking.

A consistent goal of the *Mir* program has been to maintain a permanent human presence in space. Except for two brief periods (July 1986–February 1987; April–September 1989), Russian cosmonauts have lived aboard *Mir* continuously since launch. In the process they have set several records for extended space flight. For example, Dr. Valeri Polyakov arrived on *Mir* on *Soyuz-TM 18* in January 1994 and returned to Earth on *Soyuz-TM 20* on 21 March 1995, a record of more than 438 days.

In the redesigned international space station, *Mir* was incorporated as the first part of the larger structure. In preparation for construction, the space programs of the United States and Russia met in Earth-orbit in the summer of 1995 for the first time since 1975, when the space shuttle *Atlantis* docked with the Russian *Mir*. This mission was the first of nine planned shuttle/*Mir* link-ups between 1995 and 1998, all intended to pave the way toward assembly of an international space station to be constructed in orbit beginning in November 1998. *Atlantis* lifted off on 27 June 1995 from Kennedy Space Center's Launch Complex 39-A, and docked with *Mir* on 29 June. After ceremonies following the rendezvous and docking, the two groups of spacefarers undertook several days of joint scientific investigations inside the *Spacelab* module tucked in *Atlantis*'s large cargo bay. At the end of joint docked activities on 4 July 1995, two Russian cosmonauts and American astronaut Norman H. Thagard—all of whom had been aboard the station since 16 March 1995—joined the shuttle crew for a return trip to Earth. Thagard returned home with the American record for a single space flight, with more than 100 days in space. *Atlantis* returned to the Kennedy Space Center on 7 July.

Although there is much yet to do, the seven docking missions conducted through 1997 aimed toward increasing international spaceflight capabilities seem to signal a major change in the history of space exploration. With the launch of the first international space station components in late 1998, it may well be that international competition has been replaced with cooperation as the primary factor behind huge expenditures for space operations.

Space policy analyst John M. Logsdon noted that it is remarkable that the space station program has survived to this point, because of its weak support over the years, both internationally and domestically. He added:

> One hopes that all of this is behind us now, and that for the next seven years the 14-state station partnership can focus all of its energies on finally putting together the orbital facility, without the diversion of continuing political arguments over its basic existence and overall character. . . .
>
> Even with all its difficulties and compromises, the space station partnership still stands as the most likely model for future human activities in space. The complex multilateral mechanisms for managing station operations and utilization will become a *de facto* world space agency for human space flight operations, and planning for future missions beyond Earth orbit is most likely to occur within the political framework of the station partnership.[9]

Several hundred years from now, humans may look back on the building of an international space station as tangible evidence of the beginning of a cooperative effort that was successful in creating a permanent presence for Earthlings beyond the planet. Alternatively, the station might prove to be only a minor respite in the competition between nations for economic and political supremacy.

CONCLUSION

The dream of a permanent presence in space, made sustainable by a vehicle providing routine access at an affordable price, has driven space exploration advocates since the beginning of the twentieth century. All spacefaring nations of the world have accepted that paradigm as the raison d'être of its programs in the late twentieth century. It drove the United States to develop the space shuttle as a means of achieving routine access, and it prompted an international consortium of 14 nations to build a space station to achieve a permanent presence in space. Only through the achievement of these goals, space advocates insist, will a vision of space exploration be ultimately realized that includes people venturing into the unknown. This scenario makes sense if one is interested in developing an expansive space exploration effort.

At the same time, this vision has not often been consistent with political reality in the United States. Numerous questions abound concerning the need for aggressive exploration of the solar system and the desirability of colonization on other worlds. A vision of aggressive space exploration, wrote political scientist Dwayne A. Day,

> implies that a long range human space plan is necessary for the nation without justifying that belief. Political decision-makers have rarely agreed with the view that a long range plan for the human exploration of space is as necessary as—say—a long range plan for attacking poverty or developing a strategic deterrent. Space is not viewed by many politicians as a "problem" but as at best an opportunity and at worst a luxury.[10]

Most important, the high cost of conducting space exploration almost always heads any discussion of the endeavor.

Of course, there are other, less ambitious visions of space flight that are more easily justified within the democratic process of the United States. Aimed at incremental advances, these include robotic planetary exploration and even limited space activities by humans. Most of what is presently under way within the umbrella of NASA in the United States and other nations' space agencies fall into this category. Yet these moderately ambitious space efforts also raise important questions about public policy priorities and other fiscal responsibilities. At present, however, the NASA budget stands at only about one-half of 1 percent of the federal budget, and it is declining both in real terms and as a percentage of the federal budget every year as the new century approaches. Is that too much to pay to achieve the long-sought dream of discovery and exploration of the universe?

Space has always excited and inspired humanity, just as exploration of the world beyond Europe in the fifteenth and sixteenth centuries inspired and excited. Like those earlier explorations, space holds the allure of discovery of a vast unknown awaiting human assimilation. This is particularly appealing to a society such as the United States, which has been heavily influenced by territorial expansion. In many respects, space exploration represents what the star of the 1960s television phenomenon *Star Trek* dubbed it: the "final frontier." The goals of a permanent presence in space, and of routine access to that presence, could be regarded by advocates of exploration as major steps in opening that "final frontier."

NOTES

1. George M. Low to Dale D. Myers, "Space Shuttle Objectives," 27 January 1970, George M. Low Collection, NASA Historical Reference Collection, History Office, NASA Headquarters, Washington, DC.

2. Bill Nelson, with Jamie Buckingham, *Mission: An American Congressman's Voyage to Space* (New York: Harcourt, Brace, Jovanovich, 1988), p. 296.

3. William P. Rogers et al., *Report of the Presidential Commission on the Space Shuttle Challenger Accident*, 5 vols. (Washington, DC: U.S. Government Printing Office, 1986), 2: F-1.

4. Norman R. Augustine et al., *Report of the Advisory Committee on the Future of U.S. Space Programs* (Washington, DC: U.S. Government Printing Office, 1990), p. 32.

5. "State of the Union Message, January 25, 1984," *Public Papers of the Presidents of the United States: Ronald Reagan, 1984* (Washington, DC: U.S. Government Printing Office, 1986), p. 90.

6. Quoted in Howard E. McCurdy, *The Space Station Decision: Incremental Politics and Technological Choice* (Baltimore, MD: Johns Hopkins University Press, 1990), p. 171.

7. Quoted in *ibid.*, p. 194.

8. "*Freedom* Fighters Win Again: Senate Keeps Space Station," *Congressional Quarterly Weekly Report*, 12 September 1992, p. 2722.

9. John M. Logsdon, "The Space Station Is Finally Real," *Space Times: Magazine of the American Astronautical Society* 34 (November–December 1995): 23.

10. Dwayne A. Day, "The Von Braun Paradigm," *Space Times: Magazine of the American Astronautical Society* 33 (November–December 1994): 15.

Biographies: A Gallery of Space Exploration Pioneers

Edwin E. Aldrin Jr. (1930–)

Born in Montclair, New Jersey, on 20 January 1930, astronaut Edwin E. "Buzz" Aldrin attended the U.S. Military Academy at West Point, after which he entered the U.S. Air Force and received pilot training in 1951. Aldrin flew 66 combat missions in F-86s over Korea, destroying two MIG-15 aircraft. Known by his nickname "Buzz," Aldrin was also one of the most important figures in Project Apollo's successful landing of an American on the moon in 1960s.

Aldrin became an astronaut during the selection of the third group by NASA in October 1963. On 11 November 1966 he orbited aboard the *Gemini XII* spacecraft, a 4-day, 59-revolution flight that successfully concluded the Gemini program. It proved to be a fortuitous selection, because during Project Gemini Aldrin became one of the key figures working on the problem of spacecraft rendezvous in Earth or lunar orbit, and docking them together for space flight. Without solutions to such problems, Apollo could not have been successfully completed. Holding a Ph.D. in astronautics from the Massachusetts Institute of Technology, Aldrin was ideally qualified for this work, and his intellectual inclinations ensured that he carried out these tasks with enthusiasm. Systematically and laboriously, Aldrin worked to develop procedures and tools necessary to accomplish space rendezvous and docking. He was also central in devising the methods necessary to carry out the astronauts' extravehicular activities (EVAs). That, too, was critical to the successful accomplishment of Apollo.

Aldrin was chosen as a member of the three-person *Apollo 11* crew that landed on the moon on 20 July 1969, fulfilling President Kennedy's mandate to send Americans to the moon before the end of the decade. Aldrin was the second American to set foot on the lunar surface. He and *Apollo 11* commander Neil A. Armstrong spent about 20 hours on the moon's surface before returning to the orbiting *Apollo* command module. The spacecraft and the lunar explorers returned to Earth on 24 July 1969.

In 1971 Aldrin returned to the Air Force and retired a year later. He wrote two important books about his activities in the U.S. space program. In *Return to Earth* (1970), Aldrin recounted the flight of *Apollo 11*. In *Men from Earth* (1989), Aldrin discussed the entire space race between the United States and the Soviet Union. He has been an important analyst of the space program since the 1960s. He lives near Los Angeles, California.

Neil A. Armstrong (1930–)

Neil Alden Armstrong was born on 5 August 1930 on his grandparents' farm near Wapakoneta, Ohio. His parents were Stephen and Viola Armstrong. Because Stephen Armstrong was an auditor for the state of Ohio, Neil grew up in several Ohio communities, including Warren, Jefferson, Ravenna, St. Marys, and Upper Sandusky, before the family settled in Wapakoneta. He developed an interest in flying at age 2 when his father took him to the National Air Races in Cleveland, Ohio. His interest intensified when he had his first airplane ride in a Ford Tri-Motor, a "Tin Goose," in Warren, Ohio, at age 6. From that time on, he claimed an intense fascination with aviation. At age 15 Armstrong began taking flying lessons at an airport north of Wapakoneta, working at various jobs in town and at the airport to earn the money for lessons in an Aeronca Champion airplane. By age 16 he had his student pilot's license; this was before passing his driver's test and receiving that license and before graduating from Blume High School in Wapakoneta in 1947.

Immediately after high school Armstrong received a scholarship from the U.S. Navy. He enrolled at Purdue University and began to study aeronautical engineering. In 1949 the Navy called him to active duty, during which he became an aviator. In 1950 he was sent to Korea, where he flew 78 combat missions from the aircraft carrier USS *Essex*.

After mustering out of the Navy in 1952, Armstrong joined the National Advisory Committee for Aeronautics (NACA). His first assignment was at NACA's Lewis Research Center near Cleveland, Ohio. For the next 17 years he worked as an engineer, test pilot, astronaut, and administrator for NACA and its successor agency, the National Aeronautics and Space Administration (NASA).

In the mid-1950s Armstrong transferred to NASA's Flight Research Center in Edwards, California, where he became a research pilot on many pioneering high-speed aircraft—including the well-known X-15, which is capable of achieving a speed of 4,000 mph. He flew over 200 different models of aircraft, including jets, rockets, helicopters, and gliders. He also pursued graduate studies and received a M.S. degree in aerospace engineering from the University of Southern California.

Armstrong transferred to astronaut status in 1962, one of nine NASA astronauts in the second class to be chosen. He moved to El Lago, Texas, near Houston's Manned Spacecraft Center, to begin his astronaut training. There he underwent four years of intensive training for the Apollo program to land an American on the moon before the end of the decade.

On 16 March 1966, Armstrong flew his first space mission as command pilot of *Gemini VIII* with David Scott. During that mission Armstrong piloted the *Gemini VIII* spacecraft to a successful docking with an Agena target spacecraft already in orbit. Although the docking went smoothly and the two craft orbited together, they began to pitch and roll wildly. Armstrong was able to undock the *Gemini* and used retro rockets to regain control of his craft, but the astronauts had to make an emergency landing in the Pacific Ocean.

As spacecraft commander for *Apollo 11*, the first piloted lunar landing mission, Armstrong gained the distinction of being the first person to land on the moon and the first to step on its surface. On 16 July 1969, Armstrong, Michael Collins, and Edwin E. "Buzz" Aldrin began their trip to the moon. Collins was the command module pilot and navigator for the mission. Aldrin, a systems expert, was the lunar module pilot and became the second man to walk on the moon. As commander of *Apollo 11*, Armstrong piloted the lunar module to a safe landing on the moon's surface. On 20 July 1969, at 10:56 P.M. EDT, Neil Armstrong stepped down onto the moon and made his famous statement, "That's one small step for [a] man, one giant leap for mankind." Armstrong and Aldrin spent about two and one-half hours walking on the moon collecting samples, doing experiments, and taking photographs. On 24 July 1969, the module carrying the three men splashed down in the Pacific Ocean. They were picked up by the aircraft carrier USS *Hornet.*

The three *Apollo 11* astronauts were honored with a ticker-tape parade in New York City soon after returning to Earth. Armstrong received the Medal of Freedom, the highest award offered to a U.S. civilian. Armstrong's other awards following the *Apollo 11* mission included the NASA Distinguished Service Medal, the NASA Exceptional Service Medal, 17 medals from other countries, and the Congressional Space Medal of Honor.

Armstrong subsequently held the position of deputy associate administrator for aeronautics, NASA Headquarters, Washington, D.C., in the early 1970s. In that position he was responsible for the coordination and management of overall NASA research and technology work related to aeronautics.

After resigning from NASA in 1971, he became a professor of aerospace engineering at the University of Cincinnati from 1971 to 1979. During the period of 1982–1992, Armstrong served as chairman of computing technologies for Aviation, Inc., in Charlottesville, Virginia. He then became chairman of the board of AIL Systems, Inc., an electronics systems company in Deer Park, New York. At the present time, Armstrong lives on a farm in Lebanon, Ohio.

Guion S. Bluford Jr. (1942–)

Guion Steward Bluford, born on 22 November 1942 in Philadelphia, Pennsylvania, was the first African-American astronaut in space. Chosen from the astronaut class of 1978, Bluford had previously completed a B.S. in aerospace engineering at Pennsylvania State University, 1960; M.S. in aerospace engineering, 1974, and Ph.D. in aerospace engineering with a minor in laser physics, 1975, both from the Air Force Institute of Technology. Bluford first flew on the space shuttle *Challenger* (STS-8) launched on 30 August 1983. This flight, and his role on it, signaled a new era for space exploration by the United States.

Bluford's technical assignments as a member of the NASA astronaut corps included working with the Remote Manipulator System (RMS), Spacelab-3 experiments, and space shuttle systems; verifying flight software in the Shuttle Avionics Integration Laboratory (SAIL) and the Flight Systems Laboratory (FSL); and serving as the Astronaut Office point of contact for generic Spacelab and Shuttle External Tank issues. A veteran of four space flights, Bluford was a mission specialist on STS-8 in 1983, STS-61A in 1985, STS-39 in 1991, and STS-53 in 1992.

Bluford's first mission was STS-8, which launched from the Kennedy Space Center, Florida, on 30 August 1983. This was the third flight for the space shuttle *Challenger* and the first mission with a night launch and night landing. During the mission the STS-8 crew deployed the Indian National Satellite (INSAT-1B); operated the Canadian-built RMS with the Payload Flight Test Article (PFTA); operated the Continuous Flow Electrophoresis System (CFES) with live cell samples; conducted medical experiments to understand the biophysiological effects of space flight; and activated various Earth resources and space science experiments along with four "Getaway Special" canisters. STS-8 completed 98 orbits of the Earth in 145 hours before landing at Edwards Air Force Base, California, on 5 September 1983.

Thereafter, Bluford flew on three other shuttle missions, undertaking a variety of space science and technology activities. Most recently he flew as a mission specialist on the crew of *Discovery* (STS-53), a Department of Defense mission launched on 2 December 1992 from Kennedy Space Center. Among its secondary payloads were experiments on the effects of microgravity on cells from bone tissue, muscles, and blood; and the release of 2-, 4-, and 6-inch metal spheres into space to test ground-based capabilities of detecting potentially dangerous debris in low-Earth orbit. The shuttle *Discovery* returned to Earth on 9 December 1992, landing safely at Edwards Air Force Base in California.

Bluford left NASA in July 1993 to become vice president and general manager of the Engineering Division of aerospace firm NYMA, Inc., in Greenbelt, Maryland.

Frank Borman (1928–)

Born in Gary, Indiana, on 14 March 1928, Frank Borman graduated from the U.S. Military Academy at West Point in 1950 and entered the Air Force, where he became a fighter pilot. From 1951 to 1956 he was assigned to various fighter squadrons. After completing a M.S. in aeronautical engineering, in 1957 he became an instructor of thermodynamics and fluid mechanics at West Point.

On 17 September 1962, Borman became an astronaut with the National Aeronautics and Space Administration (NASA). He commanded the *Gemini VII* mission launched on 4 December 1965, where he participated in the longest space flight to that time (330 hours and 35 minutes) and the first rendezvous of two maneuverable spacecraft.

Borman's most significant space mission was as commander of the *Apollo 8* mission, which flew around the moon over the Christmas holidays of December 1968. Initially it was planned as a mission to test Apollo hardware in low-Earth orbit, but NASA officials took a calculated risk to expand it into a circumlunar flight. As *Apollo 8* traveled outward, the crew focused a portable television camera on Earth and for the first time humanity saw its home from afar. Some people have suggested that the modern environmental movement was aided in its effort by these images of a fragile planet surrounded by total bleakness. When *Apollo 8* arrived at the moon on Christmas Eve, this image of Earth was even more strongly reinforced when Borman sent back images of the planet while reading the first part of the Bible—"God created the heavens and the Earth, and the Earth was without form and void"—before sending Christmas greetings. The flight returned to Earth on 27 December 1968. An important accomplishment, this flight united the nation, coming as it did at a time when American society was in crisis, over Vietnam, race relations, and social unrest.

Borman left NASA at the conclusion of the Apollo program and in

1975 was appointed president of Eastern Airlines. At that time the airline was operating under significant deficits. Borman restructured the airline, changing its short-haul route structure to one that emphasized continental flights. He also modernized the Eastern fleet, replacing its aging, fuel-inefficient jets with newer widebodies. These efforts brought prosperity to Eastern, and between 1976 and 1980 it enjoyed its most profitable period in history. Borman was less successful in dealing with the airline workers' unions, which had been at odds with Eastern's management for years, and with the effects of airline deregulation. Borman left Eastern Airlines in 1986 and began private consulting work from his home in Las Cruces, New Mexico.

The *Challenger* Shuttle Crew

Francis Scobee, commander
Michael Smith, pilot
Judith Resnik, mission specialist
Ronald McNair, mission specialist
Ellison Onizuka, mission specialist
Gregory Jarvis, payload specialist
Christa McAuliffe, schoolteacher

These seven astronauts died in the explosion of their spacecraft during the launch of STS-51L from the Kennedy Space Center at approximately 11:40 A.M. EST on 28 January 1986. The explosion occurred 73 seconds into the flight as a result of a leak in one of two solid rocket boosters that ignited the main liquid fuel tank. The crewmembers of the *Challenger* represented a cross-section of the American population in terms of race, gender, geography, background, and religion. The explosion became one of the most significant events of the 1980s, as millions around the world saw the accident on television and mourned the crewmembers killed.

The spacecraft commander was Francis R. "Dick" Scobee, born on 19 May 1939 in Cle Elum, Washington. After graduation from public high school in Auburn, Washington, in 1957 Scobee enlisted in the U.S. Air Force. He was trained as a reciprocating engine mechanic but longed to fly. He took night courses and in 1965 completed a B.S. in aerospace engineering from the University of Arizona. This made it possible for Scobee to receive an officer's commission and enter the Air Force pilot training program. He received his pilot's wings in 1966 and began a series of flying assignments with the Air Force, including a combat tour in Vietnam. He attended the USAF Aerospace Research Pilot School at Edwards Air Force Base, California, in 1972 and thereafter was involved in several test programs. As an Air Force test pilot, Scobee flew more than 45 types of aircraft and logged more than 6,500 hours of flight time.

In 1978 Scobee entered NASA's astronaut corps and was the pilot of

STS-41C, the fifth orbital flight of the *Challenger* spacecraft, launching from Kennedy Space Center, Florida, on 6 April 1984. During this seven-day mission the crew successfully retrieved and repaired the ailing Solar Maximum Satellite and returned it to orbit. This was an enormously important mission, because it demonstrated the capability that NASA had long said existed for the space shuttle to repair satellites in orbit.

The pilot for the fatal 1986 *Challenger* mission was Michael J. Smith, born on 30 April 1945 in Beaufort, North Carolina. At the time of the *Challenger* accident, he was a commander in the U.S. Navy. Smith had been educated at the U.S. Naval Academy, class of 1967, and received a M.S. in aeronautical engineering from the Naval Postgraduate School in 1968. Then he underwent aviator training at Kingsville, Texas, and received his wings in May 1969. After a tour as an instructor at the Navy's Advanced Jet Training Command between 1969 and 1971, Smith flew A-6 Intruders from the USS *Kitty Hawk* in Southeast Asia. Later he worked as a test pilot for the Navy, flying 28 different types of aircraft and logging more than 4,300 hours of flying time. Smith was selected as a NASA astronaut in May 1980; a year later, after completing further training, he received an assignment as a space shuttle pilot, the position he occupied aboard *Challenger*. This mission was his first space flight.

Judith A. Resnik was one of three mission specialists on *Challenger*. She was born on 5 April 1949 in Akron, Ohio. Resnik was educated in public schools before attending Carnegie-Mellon University, where she received a B.S. in electrical engineering in 1970, and the University of Maryland, where she earned a Ph.D. in the same field in 1977. Resnik worked in a variety of professional positions with the RCA Corporation in the early 1970s and as a staff fellow with the Laboratory of Neurophysiology at the National Institutes of Health in Bethesda, Maryland, between 1974 and 1977.

Selected as a NASA astronaut in January 1978, as part of the first cadre containing women, Resnik underwent the training program for shuttle mission specialists during the next year. Thereafter she filled a number of positions within NASA at the Johnson Space Center, working on aspects of the shuttle program. Resnik became the second American woman in orbit during the maiden flight of *Discovery* (STS-41D) between 30 August and 5 September 1984. During this mission she helped deploy three satellites into orbit; she was also involved in biomedical research during the mission. Afterward, she began intensive training for the STS-51L mission on which she was killed.

Ronald E. McNair was the second of three mission specialists aboard *Challenger*. Born on 21 October 1950 in Lake City, South Carolina, McNair was the son of Carl C. McNair Sr. and Pearl M. McNair. He achieved early success in segregated public schools as both a student and an athlete. Valedictorian of his high school class, he attended North Carolina A&T State University, where in 1971 he received a B.S. in physics. He went on to study physics at MIT, where he specialized in quantum electronics and laser technology. He completed his Ph.D. in 1977. As a student he performed some of the earliest work on chemical and high pressure lasers, publishing pathbreaking scientific papers on the subject.

McNair was also a physical fitness advocate and pursued athletic training from an early age. He was a leader in track and football at his high school. He also earned a black belt in Karate and during graduate school began offering Karate classes at St. Paul's AME Church in Cambridge, Massachusetts. He also participated in Karate tournaments, winning more than 30 trophies. While involved in these activities, McNair met and married Cheryl B. Moore of Brooklyn, New York, and they later had two children. After receiving his Ph.D., he began working as a physicist at the Optical Physics Department of Hughes Research Laboratories in Malibu, California, and conducted research on electro-optic laser modulation for satellite-to-satellite space communications.

This research brought McNair into close contact with the space program for the first time, and when the opportunity came he applied for astronaut training. In January 1978, NASA selected him to enter the astronaut cadre as one of the first three African Americans selected. McNair became the second African American in space between 3 and 11 February 1984, by flying on the *Challenger* mission STS-41B. During this mission McNair operated the maneuverable arm built by Canadians for moving payloads in space. The 1986 mission on which he was killed was his second shuttle flight.

Ellison S. Onizuka was the third of the mission specialists. He had been born in Kealakekua, Kona, Hawaii, on 24 June 1946 to Japanese-American parents. He attended the University of Colorado, receiving B.S. and M.S. degrees in engineering in June and December 1969, respectively. While at the university he married Lorna Leido Yoshida of Hawaii, and the couple eventually had two children. He also participated in the Air Force ROTC program, leading to a commission in January 1970. Onizuka served on active duty with the Air Force until January 1978, when he was selected as a NASA astronaut. With the Air Force in the early 1970s, he was an aerospace flight test engineer at the Sacramento Air Logistics Center. After July 1975 he was assigned to the Air Force Flight Test Center at Edwards Air Force Base, California, as squadron flight test officer and later as chief of the engineering support section.

When Onizuka was selected for the astronaut corps, he entered a one-year training program and then became eligible for assignment as a mission specialist on future space shuttle flights. He worked on orbiter test and checkout teams and launch support crews at the Kennedy Space Center for the first two shuttle missions. Because he was an Air Force officer on detached duty with NASA, Onizuka was a logical choice to serve on the first dedicated Department of Defense classified mission. He was a mission specialist on STS-51C, taking place on 24–27 January 1985 on the *Discovery* orbiter. The *Challenger* flight was his second shuttle mission.

The other two members of the *Challenger* crew were not federal government employees. Gregory B. Jarvis, a payload specialist, worked for the Hughes Aircraft Corporation's Space and Communications Group in Los Angeles, California, and had been recommended for the *Challenger* flight by his company. Jarvis was born on 24 August 1944 in Detroit, Michigan. He was educated at the State University of New York at Buffalo, where he received a B.S. in electrical engineering (1967); at Northeastern University, Boston, where he received a M.S. in the same field (1969); and at West Coast University, Los Angeles, where he completed coursework for a M.S. in management science (1973). Jarvis began work at Hughes in 1973 and served in a variety of technical positions until 1984, when he was accepted into the astronaut program under Hughes's sponsorship (after competing against 600 other Hughes employees for the opportunity). Jarvis's duties on the *Challenger* mission were to gather new information on the design of liquid-fueled rockets.

The seventh member of the crew was Sharon Christa McAuliffe, the first teacher to fly in space. She was selected from among more than 11,000 applicants from the education profession for entrance into the astronaut ranks. McAuliffe was born on 2 September 1948, the oldest child of Edward and Grace Corrigan. As a youth growing up in Framingham, a suburb of Boston, she registered excitement over the Apollo moon landing program, and wrote years later on her astronaut application form that "I watched the space age being born and I would like to participate."

McAuliffe attended Framingham State College and graduated in 1970. In the same year she married her longtime boyfriend, Steven McAuliffe, and they moved to the Washington, D.C., metropolitan area so Steven could attend Georgetown University Law School. She took a job teaching in the secondary schools, specializing in American history and social studies. The couple remained in the Washington, D.C., area for the next eight years; Christa taught and completed a M.A. from Bowie State University in Maryland. They moved to Concord, New Hampshire, in 1978 when Steven accepted a job as assistant to the state attorney general. Christa took a teaching post at Concord High School in 1982, and in 1984 she learned about NASA's efforts to locate an educator to fly on the shuttle. The intent was to find a gifted teacher who could communicate with students from space.

NASA selected McAuliffe for this position in the summer of 1984, and in the fall she took a year-long leave of absence from teaching (during which time NASA paid her salary) and trained for an early 1986 shuttle mission. She had an immediate rapport with the media, and the "teacher in space" program received tremendous popular attention as a result. It was partly because of the excitement over McAuliffe's presence on the *Challenger* that the accident had such a significant impact on the nation.

Hugh L. Dryden (1898–1965)

Throughout a long federal career, Hugh Latimer Dryden placed a unique imprint on the development of aerospace technology in the United States. He served as associate director for aeronautics of the National Bureau of Standards, 1918–1947; director of the National Advisory Committee for Aeronautics (NACA) from 1947 until the creation of the National Aeronautics and Space Administration (NASA) in 1958; and deputy administrator of NASA when it was created in response to the Soviets' success with *Sputnik*.

Born on 2 July 1898 in Pocomoke City, Maryland, the son of Samuel Isaac and Nova Hill Culver Dryden, Dryden was reared in Baltimore, where he attended the public schools and graduated with honors. Dryden attended Johns Hopkins University and completed the four-year bachelor of arts course in three years, again graduating with honors in 1918.

Influenced by Dr. Joseph S. Ames, who served for many years as chairman of NACA and who himself was a pioneer in aerodynamics, Dryden undertook a study of fluid dynamics at the Bureau of Standards while continuing his courses at the Johns Hopkins University Graduate School. His laboratory work was accepted by the university when it granted him a Ph.D. in 1919.

Dryden was promoted in 1920 to head the Bureau's Aerodynamics Section. In 1924, collaborating with Dr. Lyman J. Briggs, he made some of the earliest studies of airfoil characteristics near the speed of sound. With A. M. Kuethe, in 1929 he published the first of a series of papers on the measurement of turbulence in wind tunnels and on the mechanics of boundary layer flow. Dryden gained a reputation as a leading aeronautical scientist with several other studies in turbulence and control of the boundary layer. When he was selected to deliver the 1938 Wright Brothers lecture before the Institute of the Aeronautical Sciences (the first American to be so honored), he chose the subject "Turbulence and the Boundary Layer," a critical issue in flight aerodynamics at the time.

During World War II he served on several technical groups advising the armed forces on aeronautical matters and guided missiles. As head of a project for the National Defense Research Committee, he led development of this country's first guided missile to be successfully used in combat, the radar-homing *Bat*. This achievement won him the Presidential Certificate of Merit in 1948. He also served on other committees, advising the Joint Chiefs of Staff, the NACA, the Army Ordnance Department, and the Army Air Forces on the subject of guided missiles. Following the end of the war, he continued his interest in the Bureau's guided missile development programs.

In 1945 Dryden was made deputy scientific director of the Army Air Forces (AAF) Scientific Advisory Group, which was to prepare a report as a guide for future AAF research and development programs. With this group he traveled to Germany, France, England, and Switzerland to study foreign scientific efforts in the development of aeronautics and aerial weapons, especially guided missiles. There he became acquainted with the significant efforts of Germany during the war in jet and rocket weaponry, and his reports for the AAF emphasized the need for the United States to invest in these arenas for postwar defense.

While he was completing these studies, in 1947 Dryden resigned from the Bureau of Standards to become director of aeronautical research at the NACA. Two years later the agency gave him added responsibilities and the new title of director. In this capacity he had charge of an expanding research organization with some 8,000 employees, three large laboratories, and two smaller research stations.

During his tenure, Dryden wrote 17 technical reports for the NACA relating directly to his research in aerodynamics. Additionally, the results of his work appeared in several professional and trade journals. All these dealt with the properties of airfoils at high speeds, wind-tunnel investigations, boundary layer and turbulence, noise suppression, and other aeronautical matters. He also served as editor of the *Journal of the Institute of the Aeronautical Sciences* from 1941 to 1956.

More important, during his tenure as NACA director, Dryden guided the organization into pivotal research and development in high-speed flight and rocketry. He fostered the pathbreaking research projects of the X-1, which flew faster than the speed of sound in 1947, and the X-15 hypersonic research vehicle of the 1950s and 1960s. He also opened the door for rocketry research by supporting the efforts of the Space Task Group at the Langley Aeronautical Laboratory, Hampton, Virginia, in the 1950s.

At the time of NASA's creation in 1958 (and the incorporating of the old NACA into that new organization), Dryden became the deputy administrator for T. Keith Glennan, President Eisenhower's appointee for the top NASA position. In that capacity Dryden handled the day-to-day operations of the agency and oversaw its technical efforts. Dryden threw himself into the intricacies of space flight with even greater zeal than before. The conception and planning of Project Mercury, for instance, bore his mark from the very beginning, because it emphasized the scientific component. As a result the program was neither as spectacular nor as swift as those who wanted to race the Soviet Union into space would have liked, but it yielded more knowledge about the rigors of space flight because of Dryden's leadership.

When the Kennedy administration designated James E. Webb for NASA administrator in 1961, Dryden stayed on as his deputy and provided stability in the organization, which was rapidly changing to carry out Project Apollo. Dryden found it possible to make such contributions because he and Webb had established clearly defined spheres of operation. Webb's highest priority was to lobby and win the support of Washington's political elite for the space program as well as to convince the voting public of its value. The two men shared responsibility for the broad policy direction of the agency, but Dryden had to make the hardest decisions involving technical and fiscal choices: which systems and subsystems to fund or eliminate, to accept as presented or to modify; which scientific experiments to pursue; how to structure programs for maximum usefulness; how to obtain the cooperation of universities, corporations, and foreign powers; and how to prepare and present budgets to congressional committees.

Dryden worked diligently at NASA throughout the first half of the 1960s, serving as deputy administrator until his death on 2 December 1965. His quiet oversight of the agency helped immeasurably in keeping it on track; his death was a blow both to the dynamism of the agency and to the conduct of Project Apollo. Robert C. Seamans Jr., the "third man" in the NASA leadership during this period, believed that if Dryden had lived longer his watchfulness might have foreseen and prevented the disaster of the *Apollo 204* capsule fire in January 1967.

Although many of Dryden's colleagues at NASA knew he was a religious man, following his death they learned just how intense his religious convictions had been. Numerous eulogies explained how he had incorporated religion into his daily life. He had long been active in the work of the Men's Bible Class of the Calvary Methodist Church in Washington, D.C. Moreover, he had held a Methodist preacher's license since his college days and had engaged in significant Christian activity throughout his life.

James C. Fletcher (1919–1991)

James Chipman Fletcher was born on 5 June 1919 in Millburn, New Jersey. He received an undergraduate degree in physics from Columbia University and earned a doctorate in physics from the California Institute of Technology in 1939. After holding research and teaching positions at Harvard and Princeton universities, he joined Hughes Aircraft in 1948 and later worked at the Guided Missile Division of the Ramo-Wooldridge Corporation in Los Angeles, California.

In 1958 Fletcher co-founded the Space Electronics Corporation in Glendale, California; the company became the Space General Corporation after a merger. He was later named systems vice president of Aerojet

General Corporation in Sacramento, California. In 1964 he became president of the University of Utah, a position he held until he was named NASA administrator in 1971.

After becoming NASA administrator in 1971, Fletcher succeeded in gaining the approval of the Nixon administration on 5 January 1972 to develop the space shuttle as the follow-on human spaceflight effort of the agency. He left the post with the change of presidential administrations in 1977 to return to private industry. He also served as NASA administrator a second time, for nearly three years following the loss of the space shuttle *Challenger* on 28 January 1986.

During his first administration at NASA, Fletcher was responsible for beginning the shuttle effort as well as the Viking program that sent landers to Mars. During his second tenure, he presided over the effort to recover from the *Challenger* accident. Indeed, he oversaw or initiated virtually every major space project of the 1980s and 1990s. Although the missions were planned in the 1960s before he took over, he was administrator during the three Skylab missions in 1973 and 1974 and the two Viking probes that landed on Mars in 1976. He also approved the Voyager space probe to the outer planets, the Hubble Space Telescope program, and the Apollo-Soyuz mission, which in 1975 linked American astronauts and Soviet cosmonauts in space.

During his second administration at NASA, Fletcher was largely involved in efforts to recover from the *Challenger* accident. Immediately after the accident, the shuttle program was inactive for three years while NASA worked to redesign the solid rocket boosters and change its management structure. Fletcher ensured that NASA reinvested heavily in the program's safety and reliability, made organizational changes to improve efficiency, and restructured its management system. Most important, he oversaw a complete reworking of the components of the shuttle to enhance its safety and added another method for the astronauts to exit the shuttle. A critical decision resulting from the accident and its aftermath (during which the nation had a reduced capability to launch satellites) was to expand greatly the use of expandable launch vehicles. Fletcher was in charge of the agency when the space shuttle finally returned to flight on 29 September 1988.

When he left NASA the first time in 1977, Fletcher became an independent consultant in McLean, Virginia, and served on the faculty of the University of Pittsburgh. During the nine years between his terms as NASA administrator, Fletcher was extremely active as an advisor to key national leaders involved in planning space policy. Among other activities, he served on an advisory board involved in developing the Strategic Defense Initiative.

Fletcher died of lung cancer at his home in suburban Washington on 22 December 1991.

Yuri A. Gagarin (1934–1968)

Cosmonaut Yuri Aleksevich Gagarin became the first human in space with a one-orbit mission aboard the spacecraft *Vostok 1* on 12 April 1961. He was born on 9 March 1934, the son of a collective farmer (farmers were forced into collective farms by Stalin in the 1930s) in the village of Klushino in the Gzhatsk district of the Smolensk region in the Soviet Union. Like many Soviet citizens of his generation, his education was interrupted in 1941 by the German invasion of his homeland. His family fled inland from the German advance before joining in the fight that turned back the invasion.

After World War II the Gagarins returned to the town of Gzhatsk, where young Gagarin completed his basic education. From there he went to the vocational school in the town of Lyubertsy near Moscow, specializing in moulder-smelting (metal-working) and he completed his degree in 1951. He later attended the Saratov Industrial Technical School, eventually earning a diploma with honors. While at the technical school Gagarin joined a local flying club and developed aviation skills. In 1955 he joined the Orenburg Aviation College, and after completing its course in 1957 he entered the Soviet air force and became a fighter pilot. While serving in the military, Gagarin also completed studies at the Zhukov Air Force Engineering Academy.

He was chosen for cosmonaut training in 1959 and underwent a series of increasingly rigorous physical and mental exercises to prepare for space flight. Selected from among several other contenders for the first flight in space, Gagarin represented the Soviet ideal of the worker who rose through the ranks solely on the basis of merit. His handsome appearance, thoughtful intellectuality, and boyish charm made him an attractive figure on the world stage. The importance of these attributes was not lost on Nikita Khrushchev and other senior Soviet leaders.

The launch of the first Soviet orbital mission in April 1961 proved enormously important both for Gagarin and for the Soviet Union. *Vostok 1* was a three-ton ball-shaped capsule, with an attached two-ton equipment module containing (among other things) retrorockets. It was lofted into orbit atop a modified R-7 ICBM rocket. His 108-minute flight had been the direct result of Cold War competition between the United States and the Soviet Union to be the first to place a human in space, as a demonstration of technological superiority before the world. With Gagarin's success, the United States lost that challenge and the Soviet Union was recognized as a technological and scientific superpower.

After the announcement of the Soviet flight, U.S. president John F. Kennedy congratulated the Soviet Union for its success but stated that "no one is more tired than I am" of seeing the United States take second place behind the Soviets in the space field. "They secured large boosters which have led to their being first in Sputnik, and led to their first putting their men in space. We are, I hope, going to be able to carry out our efforts, with due regard to the problem of the life of the men involved, this year. But we are behind . . . the news will be worse before it is better, and it will be some time before we catch up," said Kennedy in a press conference in 1961.

Despite its outward success, Gagarin's *Vostok 1* flight had serious problems. Since the end of the Cold War, increasing amounts of information have confirmed what some analysts believed all along—namely, that Gagarin's flight had nearly been a disaster when the capsule spun dangerously out of control while beginning the re-entry sequence. Gagarin told officials during a post-flight debriefing, "As soon as the braking rocket shut off, there was a sharp jolt, and the craft began to rotate around its axis at a very high velocity." It spun uncontrollably as the equipment module failed to separate from the cosmonaut's capsule. After ten minutes the spacecraft stabilized somewhat from its dizzying spin. Nevertheless, Gagarin ejected from the capsule and parachuted safely to Earth while the capsule came down elsewhere, dangling from its own chutes.

In the aftermath Gagarin's flight and the international publicity that came from it, President Kennedy asked his vice president, Lyndon B. Johnson, to explore possibilities whereby the United States might recover its status as a first-rate spacefaring nation. On 20 April 1961 he wrote a memo to Johnson asking him to learn if "we have a chance of beating the Soviets by . . . a trip around the moon, or by a rocket to land on the moon, or by a rocket to go to the moon and back with a man. Is there any other space program which promises dramatic results in which we could win?" These questions led to a far-reaching reconsideration of space activities for the United States and Kennedy's announcement on 25 May 1961 on national television "that this nation should commit itself to achieving the goal, before this decade is out, of landing a man on the moon and returning him safely to the earth."

In the context of Yuri Gagarin's 12 April 1961 flight on *Vostok 1*, Project Apollo achieved higher priority within NASA than those who had first proposed a lunar landing mission had ever envisioned. As early as 1959, the lunar program had been a critical component of NASA's long-range planning effort. With Kennedy's support, it became not only a critical component but essentially the *only* component in the effort. It culminated between 1969 and 1972 with six landings on the moon.

Following Gagarin's triumph, on 14 April 1961 the Soviet Union held a gigantic ceremony in Red Square in Moscow honoring its first cosmonaut. The great success of the *Vostok 1* flight made the gregarious Gagarin a global hero, and he was an effective spokesman for the Soviet Union on the world stage. He died in a plane crash while on a training mission for the Soviet air force on 27 March 1968.

Gagarin's space flight energized the Soviet leadership to invest more money in space exploration during the years that followed, in part because of the international prestige that the nation gained for its spectacular missions. Subsequently the Soviets were the first to launch a woman, Valentina Tereshkova, into orbit; the first to launch two- and three-person crews, and the first to execute an extravehicular activity, or spacewalk.

John H. Glenn Jr. (1921–)

John H. Glenn Jr. served as the pilot for the 20 February 1962 Mercury-Atlas 6 (*Friendship 7*) mission, the first American orbital space flight. He made three orbits on this mission, in the process sealing his place in history.

Glenn was born on 18 July 1921 in Cambridge, Ohio. Shortly thereafter his family moved to nearby New Concord, where after graduating from New Concord High School he enrolled at Muskingum College. By that time he had already learned to fly at a small airfield in New Philadelphia. Not long after the U.S. entry into World War II, he decided to pursue aviation as a career and enlisted in the Naval Aviation Cadet Program. He was commissioned in the Marine Corps in 1943 and served in combat in the South Pacific. Glenn remained in the Marines after the war, and during the Korean conflict (1950–1953) he also flew combat missions. For his service in 149 missions during those two wars he received many honors, including the Distinguished Flying Cross (six occasions) and the Air Medal with 18 clusters.

Thereafter, Glenn served several years as a test pilot on Navy and Marine Corps jet fighters and attack aircraft, setting a transcontinental speed record in 1957 for the first flight to average supersonic speeds from Los Angeles to New York.

In 1959 John Glenn was selected to be one of the first seven astronauts in the U.S. space program. Three years later, on 20 February 1962, he made history as the first American to orbit the Earth, completing three orbits in a five-hour flight. For this achievement he received the Congressional Space Medal of Honor. He left NASA in 1964 to re-enter civilian life.

Returning to his native Ohio, in 1965 Glenn began to take an active part in politics and early environmental protection efforts while pursuing a career as an executive with Royal Crown International, a soft drink manufacturer. In 1974 he ran successfully for the U.S. Senate, carrying all 88 counties of Ohio, and was re-elected in 1980 with the largest margin in Ohio history. Ohioans returned him to the Senate for a third term in 1986, again with a substantial majority. In 1992, John Glenn again made history by being the first popularly elected senator from Ohio to win four consecutive terms. In the early part of 1997 he announced that with the completion of his fourth term in the Senate he will retire and return to space. On January 16, 1998, NASA announced that Senator Glenn would fly as a payload specialist on STS-95, *Discovery*, in October 1998. On it he will participate in experiments on the physiology of aging Americans.

As a senator, Glenn was uniquely placed to aid NASA by supporting its budget initiatives for space exploration projects. Although throughout his career in the Senate he has capably explained the possibilities of movement beyond the boundaries of Earth and has been an advocate for well-developed projects, he has not been unswervingly devoted to NASA's projects. He provided much-needed political support for NASA's efforts to win funds to build the space shuttle, and he has usually supported robotic science missions to the planets; but he has also questioned the massive expenditures necessary to build a space station, and he did not support the missions to the moon and Mars proposed by the Bush administration in 1989. A moderate Democrat representing a state that did not benefit significantly from NASA expenditures, Glenn has taken a measured approach to supporting individual space projects while always providing strong and sometimes eloquent advocacy for space exploration as a long-term objective of the United States.

Robert H. Goddard (1882–1945)

The son of a machine shop owner, Robert Hutchings Goddard was born in Worcester, Massachusetts, on 5 October 1882. He pioneered modern rocketry in the United States and founded an entire field of science and engineering. Goddard graduated from Worcester Polytechnic Institute in 1908 and then became a physics instructor at Worcester Technical University, where he received a M.A. in 1910 and a Ph.D. in 1911. Goddard was a research fellow at Princeton in 1912 and 1913 and then joined the faculty at Clark University, where he became a full professor in 1919.

Motivated by reading science fiction as a boy, Goddard became excited by the possibility of exploring space. As a youth in 1901 he wrote a short paper, "The Navigation of Space," that argued that movement could take place by firing several cannons "arranged like a 'nest' of beakers." At his high school oration in 1904 he summarized his life's perspective: "It is difficult to say what is impossible, for the dream of yesterday is the hope of today and the reality of tomorrow." In 1907 he wrote another paper on the possibility of using radioactive materials to propel a rocket through interplanetary space. He sent this article to several magazines, and all rejected it.

Goddard had an especially inquisitive mind that became curious about space flight by reading and writing science fiction. For instance, as an undergraduate he described in a short story a railroad system between Boston and New York in which the trains traveled in a vacuum under the pull of an electromagnetic field and completed their trip in ten minutes.

As a young physics graduate student he conducted static tests with small solid-fuel rockets at Worcester Tech, and in 1912 he developed a detailed mathematical theory of rocket propulsion. He continued these efforts and actually received two patents in 1914. One was the first for a rocket using solid and liquid fuel, and the other for a multistage rocket. In 1915 he proved that rocket engines could produce thrust in a vacuum and therefore make space flight possible. In 1916 the Smithsonian Institution provided funds for Goddard to continue his work on solid-propellant rockets and to begin development of liquid-fuel rockets as well.

During World War I, Goddard further explored the military possibilities of rockets. He succeeded in developing several types of solid-fuel rockets to be fired from hand-held or tripod-mounted launching tubes, which formed the basis of the bazooka and other powerful rocket weapons of World War II.

After serving in World War I, Goddard became a professor of physics at Clark College (later renamed Clark University) in Worcester, Massachusetts. There he turned his attention to liquid rocket propulsion, theorizing that liquid oxygen and liquid hydrogen were the best fuels, but learning that oxygen and gasoline were less volatile and therefore more practical. To support his investigations, Goddard applied to the Smithsonian Institution for assistance in 1916 and received a $5,000 grant from its Hodgkins Fund. His research was ultimately published by the Smithsonian as the classic study *A Method of Reaching Extreme Altitudes* in 1919. Here Goddard argued from a firm theoretical base that rockets could be used to explore the upper atmosphere. Moreover, he suggested that with a velocity of 6.95 miles per second, without air resistance, an object could escape Earth's gravity and head into infinity, or toward other celestial bodies. This became known as Earth's "escape velocity." He also argued that humans could reach the moon using these techniques.

These ideas became a great joke for those who believed space flight was either impossible or impractical. Some ridiculed Goddard's ideas in the popular press, which caused the already shy Goddard to become even more so. Soon after the appearance of his publication, he commented that he had been "interviewed a number of times, and on each occasion have been as uncommunicative as possible." The *New York Times* was especially harsh in its criticisms, referring to him as a dreamer whose ideas had no scientific validity. It also compared his theories to those advanced by novelist Jules Verne, indicating that such musing is "pardonable enough in him [Verne] as a romancer, but its like is not so easily explained when made by a savant who isn't writing a novel of adventure." The *New York Times* questioned Goddard's credentials as a scientist and the Smithsonian's rationale for funding his research and publishing his results in an editorial on 18 January 1920.

Such negative publicity prompted Goddard to become even more secretive and reclusive. However, it did not stop his work, and he eventually registered 214 patents on various components of rockets. He concentrated on the design of a liquid-fueled rocket (the first such design) and the related fuel pumps, motors, and control components. On 16 March 1926 near Auburn, Massachusetts, Goddard launched his first rocket, a liquid oxygen and gasoline vehicle that rose 184 feet in 2.5 seconds. This event heralded the modern age of rocketry. He continued to experiment with rockets and fuels for the next several years. A spectacular launch took place on 17 July 1929 when he flew the first instrumented payload—an aneroid barometer, a thermometer, and a camera (to record the readings). It was the first instrument-carrying rocket. The launch failed; after rising about 90 feet the rocket turned and struck the ground 171 feet away. It caused such a fire that neighbors complained to the state fire marshal and Goddard was prohibited from making further tests in Massachusetts.

Fortunately, Charles A. Lindbergh, fresh from his transatlantic solo flight, became interested in Goddard's work. He visited Goddard and was sufficiently impressed to persuade Daniel Guggenheim, a philanthropist, to award Goddard a grant of $50,000. With this, Goddard set up an experiment station in a lonely spot near Roswell, New Mexico. Here he built larger rockets and developed many theories that are now standard in rocketry. He designed combustion chambers of the appropriate shape, and he burned gasoline with oxygen in such a way that the rapid combustion could be used to cool the chamber walls.

From 1930 to 1941 he launched rockets of increasing complexity and capability. He developed systems for steering a rocket in flight by using a rudder-like device to deflect the gaseous exhaust, with gyroscopes to keep the rocket headed in the proper direction. Goddard described many of his results in 1936 in a classic study, *Liquid-Propellant Rocket Development*. The culmination of this effort was the successful launch of a rocket to an altitude of 9,000 feet in 1941. In late 1941 Goddard entered naval service and spent the duration of World War II developing a jet-assisted takeoff (JATO) rocket to shorten the distance required for heavy aircraft launches. Some of this work led to the development of the "throttlable" Curtiss-Wright XLR25-CW-1 rocket engine, which later powered the Bell X-1 and helped overcome the transonic barrier, a range of speed between about .9 and 1.1 mach in 1947. Goddard did not live to see this; he died of cancer in Baltimore, Maryland, on 10 August 1945.

Goddard accomplished a great deal, but because of his modesty most people did not know about his achievements during his lifetime. These included theorizing on the possibilities of jet-powered aircraft, rocket-borne mail and express, passenger travel in space, nuclear-powered rockets, and journeys to the moon and other planets. He also made the first mathematical exploration of the practicality of using rockets to reach high altitudes and achieve escape velocity. He patented numerous inventions associated with space flight.

When German rocket experts were brought to America after World War II and were questioned about rocketry, they suggested talking to Goddard, the pioneer in the field. American officials could not do so because Goddard had already died and his achievements had been overlooked. In 1960 the U.S. government recognized Goddard's work when the Department of Defense and the National Aeronautics and Space Administration (NASA) awarded his estate $1 million for the use of his 214 rocketry patents. Although he did not live to see the space age begin, if any one man had a central role in its creation, it was Goddard.

Sergei P. Korolev (1906–1966)

For years the details of the life and career of Sergei Pavlovich Korolev, the chief designer of early Soviet rockets, were kept in mystery as a state secret. Born on 30 December 1906 at Zhitomir, the son of a teacher, Korolev became interested in the possibilities of space flight at a young age. Trained in aeronautical engineering at the Kiev Polytechnic Institute, in 1930 he co-founded the Moscow rocketry organization GIRD (Gruppa Isutcheniya Reaktivnovo Dvisheniya, or Group for Investigation of Reactive Motion). Like the VfR (Verein fur Raumschiffarht, or Society for Spaceship Travel) in Germany, and Robert H. Goddard in the United States, by the early 1930s the Russian organizations were testing liquid-fueled rockets of increasing size.

In Russia, GIRD lasted only two years before the military, recognizing the potential of rockets, replaced it with the RNII (Reaction Propulsion Scientific Research Institute). RNII developed a series of rocket-propelled missiles and gliders during the 1930s, culminating in Korolev's RP-318, Russia's first rocket-propelled aircraft. Before the aircraft could make a rocket-propelled flight, however, Korolev and other aerospace engineers were imprisoned during 1937–1938 at the peak of Stalin's purges. During this time of paranoia, people of competence often received sentences of death or imprisonment simply because of a perception of disloyalty. Korolev and several other rocket designers were victims of this paranoia, although there is no evidence that Korolev himself was involved in any traitorous activities. Korolev at first spent months in transit on the Transsiberian railway and on a prison vessel at Magadan. This was followed by a year in the Kolyma gold mines, the most dreaded part of the Gulag prison camp of political enemies of the Soviet Union.

However, Stalin soon recognized the importance of aeronautical engineers in preparing for the impending war with Hitler, and he released from prison Korolev and other technical personnel who could help the Red

Army by developing new weapons. A system of *sharashkas* (prison design bureaus) was set up to exploit the jailed talent. Korolev was saved by the intervention of senior aircraft designer Sergei Tupolev, himself a prisoner, who requested Korolev's services in the TsKB-39 *sharashka*. However, Korolev was not allowed to work on rockets except at night on his own time.

With victory in World War II virtually assured by 1944, and seeing the immense progress that Wernher von Braun's team had made with the V-2 rocket in Nazi Germany, Stalin decided to develop ballistic missiles of his own. He sent Korolev and other technical experts from the Kazan *sharashka* to Soviet-occupied Germany to investigate von Braun's efforts in 1945. At first Korolev merely accompanied the team that salvaged what was left of the V-2 production effort, but soon he began interviewing dozens of V-2 engineers and technicians who still remained in Germany.

On 13 May 1946, Stalin signed the decree initiating development of Soviet ballistic missiles. The minister of armaments, Dmitri Fedorovich Ustinov, was placed in charge of the development. In August 1946 the Scientific Research Institute NII-88 was established to conduct the development. Korolev's personality and organizational abilities had been impressive, and Ustinov personally appointed him chief constructor for development of a long-range ballistic missile. Following Korolev's instructions, 200 German employees of the Mittelwerke V-2 factory were rounded up on the night of 22–23 October 1946 and sent to relatively comfortable living quarters at Lake Seleger, between Moscow and Leningrad. Thus the jailed became the jailer. The Germans had little direct contact with Korolev's engineers. Aside from assisting in the launch of a few more V-2s from Kapustin Yar, they mainly answered written questions. They were finally returned to Germany between 1950 and 1954.

The V-2, initially copied with all Soviet components as the R-1, was quickly developed into successively more capable R-2 and R-5 missiles. By 1 April 1953, as Korolev was preparing for the first launch of the R-11 rocket, he received approval from the Council of Ministers for development of the world's first intercontinental ballistic missile (ICBM), the R-7. To concentrate on development of the R-7, Korolev's other projects were reassigned to a new design bureau in Dnepropetrovsk headed by Korolev's assistant, Mikhail Kuzmich Yangel. This was the first of several design bureaus (some of which later competed with Korolev's) that would be spun off once Korolev had perfected a new technology.

It was Korolev's R-7 ICBM that launched *Sputnik 1* on 4 October 1957. This launch galvanized American concern about the capability of the Soviet Union to attack the United States with nuclear weapons using ballistic missiles. Indeed, the Soviet Union's succession of Sputnik and Luna launches, combined with the belligerent claims of Premier Nikita Khrushchev, created the public impression that the Soviet Union was far ahead of the United States in the development of unstoppable ICBMs and space weapons. In fact, Korolev's R-7—with its enormous launch pads, complex assembly and launching procedures, cryogenic liquid oxygen oxidizer, and radio-controlled terminal guidance—was a thoroughly impractical weapon. The warhead was overly heavy and therefore had a reduced range of only about 3,500 miles, barely enough to reach the northern United States. As a result, it would be deployed as a weapon at only eight launch pads in Tyuratam and Plesetsk, in the northern USSR. Development of more practical successors, such as Korolev's R-9, was not begun until 13 May 1959.

Soviet leaders then asked Korolev to develop even more capable launchers, and the immediate result was the RT-2. This was a tall order. While he completed theoretical studies of the next generation of launch vehicles (the N vehicle) and spacecraft (Vostok Zh and Soyuz B), others inside the Soviet space technology bureaucracy persuaded Khrushchev in 1961 to proceed with development of the launch vehicle (UR-500 Proton) and the spacecraft (the LK-1) for a piloted circumlunar mission to follow Earth-orbital missions.

The Soviet space program of the early 1960s came to resemble the cautious personality of Sergei Korolev, who wanted definitely to explore space, but to do it safely. Because of safety concerns, Korolev made sure his designs evolved gradually over time, always using a design that worked safely and then building on the success. The Vostok capsule evolved directly into the Soyuz capsule, which underwent several subsequent design changes but is still in use. Only the Voskhod program, forced on Korolev by Khrushchev as a prestige program, was an abnormal design that substituted three cosmonaut seats for the ejection system so the Soviet Union could beat the United States in launching a three-person crew into space.

Following Voskhod, Korolev campaigned to send a Soviet cosmonaut to the moon. Following the initial reconnaissance of the moon by *Luna*s *1*, *2*, and *3*, Korolev established three largely independent efforts aimed at achieving a Soviet lunar landing before the Americans. The first objective, met by Vostok and Voskhod, was to prove that human space flight was possible. The second objective was to develop lunar vehicles that would soft-land on the moon's surface (the soft landing would ensure that the vehicle would not sink into the dust accumulated by four billion years of meteorite impacts). The third objective was to develop a huge booster to send cosmonauts to the moon.

The most difficult of these objectives was the third one. Korolev's design bureau began work on the N-1 launch vehicle, a counterpart to the American Saturn V, beginning in 1962. This rocket was to be capable of launching a maximum of 110,000 pounds into low-Earth orbit. Although the project continued until 1971 before cancellation, the N-1 never made a successful flight.

By 1964 the N-1 program was in trouble. Because of design considerations for the lunar landing craft and the orbiting command capsule, the launcher needed the capability of putting 92 metric tons into low-Earth orbit. This capability called for more main engines, and the N-1 already had 30. Getting them all to work proved more than Korolev's engineers could achieve. Also, the N-1's payload capability could only support two cosmonauts going to the moon and only one cosmonaut actually landing on the lunar surface.

Moreover, Kruschchev directed Korolev to accomplish the lunar effort—at least a circumlunar flight—by 1967 in commemoration of the fiftieth anniversary of the Bolshevik revolution. Because of this deadline, Korolev pressed his rocket design bureau to develop LH2 and LOX engines for all three N-1 stages. In October 1964 Premier Khrushchev was removed from office by a coup; this cost Korolev a strong ally at the head of the government, but it did not ease the time schedule for completion of a lunar flight. Now, instead of a relentless schedule and resources made available by an enthusiastic premier to meet it, Korolev had only a relentless schedule.

On 14 January 1966, Sergei P. Korolev died from an improperly performed hemorrhoid operation. Because of his importance in the rocketry program, the Soviet Minister of Health had insisted on performing the operation himself—and when he found tumors in Korolev's intestines, the doctor continued without help, appropriate medical supplies, or extra blood. In death, Korolev received accolades for the first time for his successes in space flight. Having been known previously in the West only as the "Chief Designer," now his true identity was revealed to the world and the Soviet Union accorded him a hero's funeral and burial in the Kremlin Wall. When Korolev died, however, any realistic possibility of beating the Americans to the moon also died. Several of Korolev's lieutenants and rivals emerged to direct what was left of the lunar landing program, but political intrigue and technical failures forced its eventual cancellation.

Hermann Oberth (1894–1989)

One of the most significant rocketry pioneers of the twentieth century was Hermann Oberth, by birth a Romanian but by nationality a German. Born on 25 June 1894 in Hermannstadt, Romania, as an 11-year-old Oberth became mesmerized by Jules Verne's novel, *From Earth to the Moon.* He recalled reading the book "five or six times and, finally, knew it by heart." This book, and other spaceflight literature that he devoured in the coming years, led Oberth to intensive study of the technical aspects of interplanetary travel.

Although he studied for a career in medicine, Oberth never could shake his obsession with space flight and finally switched his emphasis to physics. He wrote a dissertation on the problem of rocket-powered flight, but his work was rejected by the University of Heidelberg in 1922 for being too speculative. However, this dissertation became the basis for his classic 1923 book, *Die Rakete zu den Planetenräumen (The Rocket into Interplanetary Space)*. The book explained the mathematical theory of rocketry, applied it to possible designs for practical rockets, and considered the potential of space stations and human travel to other planets.

The success of the 1923 book prompted Oberth to consider writing a more popular, and less technical, treatise on the possibilities of space flight, but because of his teaching load in a secondary school, German spaceflight enthusiast Max Valier condensed and published one for him. This book inspired a number of new rocket clubs to spring up all over Germany, as hardcore rocket enthusiasts tried to translate Oberth's theories into practical space vehicles. The most important was the Verein für Raumschiffarht (Society for Spaceship Travel), or VfR. Oberth became a sort of mentor for the VfR during the 1920s, encouraging the efforts of Valier, Willy Ley, and the young Wernher von Braun.

In 1929 Oberth published another major work, *Wege Zur Raumschiffahrt (The Road to Space Travel)*, in which he envisioned the development of ion propulsion and electric rockets. This book won an award established by the French rocket pioneer Robert Esnault-Pelterie, and Oberth used the prize money to buy rocket motors for the VfR.

One man who foresaw the possible outcomes of the space program was the silent movie maker Fritz Lang. After reading Oberth's book, he decided to film an adventure story about space travel. The result was the 1929 feature *Die Frau Im Mond (The Woman in the Moon)*. Lang wanted his movie set to be technically correct, so he asked Hermann Oberth to be his main technical advisor. Oberth and Willy Ley helped Lang with his sets and built a spacecraft that looked very realistic. Ever the dramatist, Lang even invented the countdown to increase tension for the audience and to add drama to the rocket flight.

As a publicity stunt for Lang's film, Oberth also agreed to build an actual rocket that would be launched at the premier of *Die Frau Im Mond*. Two days before the premier, however, Oberth discovered that the rocket would not be completed in time. At that point he went to Romania to soothe his nerves. After 1938 Oberth was involved in a series of research projects concerning rockets for Germany. In 1941 he became a naturalized German citizen, and during World War II he worked for Wernher von Braun in the V-2 development program—but never held an important position in the project. At the end of the war, Oberth was interrogated by American captors and then released. He settled in Feucht, West Germany, near Nuremberg.

In 1955 Wernher von Braun, by then the head of a U.S. Army ballistic missile effort at Huntsville, Alabama, invited Oberth to work for him on his program. Oberth worked for a short time on these efforts, but in 1959 he retired and returned to Feucht, where he spent the rest of his life. Because of his significance as the "godfather" of early German rocketry, Oberth returned to the United States in July 1969 to witness the launch of the Saturn V rocket that carried the *Apollo 11* crew on the first lunar landing mission. He then returned to Germany, where he died on 29 December 1989, having helped to create and sustain space flight and witness many of the major events of space exploration in the latter half of the twentieth century.

Sally K. Ride (1951–)

Sally Kristen Ride was the first American woman to fly in space. Born on 26 May 1951 in Los Angeles, California, she is the daughter of Dale B. Ride, a professor at Santa Monica Community College, and Joyce Ride, a counselor at a women's correctional institution. Sally began early to combine her competitive spirit with academic determination. As a youngster, her ability on the tennis courts led her to rate eighteenth nationally on the junior tennis circuit. Initially studying physics at Swarthmore College, she finished at Stanford University, earning her B.S. in physics and B.A. in English literature in 1973. Ride received her Ph.D. in physics at Stanford University in 1978.

Hearing that the National Aeronautics and Space Administration (NASA) was looking for young scientists to serve as mission specialists, she applied and was selected in the astronaut class of 1978—the first in which women were permitted entrance. She made her historic flight in 1983 aboard STS-7, the flight that set the pattern for combined-gender shuttle flights. As part of the seventh flight of the space shuttle *Challenger* (STS-7) on 18 June 1983, this mission launched two commercial satellites (Anik C-2 and Palapa B-1) and deployed and retrieved SPAS 01, a scientific satellite.

After the *Challenger* accident in January 1986, Ride was appointed to serve as a member of the Presidential Commission on the Space Shuttle *Challenger* Accident (the Rogers Commission) in 1986. In 1986–1987 she also chaired a NASA task force that prepared a report on the future of the civilian space program. Entitled *Leadership and America's Future in Space*, this report served as a blueprint for restructuring the agency after the accident and helping it return to space flight.

Ride resigned from NASA in 1987 to join the Center for International Security and Arms Control at Stanford University. She left Stanford in 1989 to assume her current positions as director of the California Space

Institute, part of the University of California at San Diego, and as professor of physics at the university. Ride remains active in her scientific pursuits and as a member of several scientific advisory bodies for various organizations of the federal government.

Alan B. Shepard Jr. (1923–)

Alan Bartlett Shepard Jr. was born on 18 November 1923 in East Derry, New Hampshire. He was a career Navy pilot before being chosen as one of the first seven astronauts by NASA in 1959. The first American to ride a rocket into space, he represented a technological revolution that began seriously in the 1960s: the new great age of discovery, the exploration of areas where humans cannot live without artificial habitations—in space, under the ocean, and at the poles.

Shepard's role as the first American astronaut made him a celebrity, fueled by the American people's interest in human space flight and shaped by NASA's skillful public relations. In the popular imagination he was a frontiersman in the same mold as Lewis and Clark. His actions as one of seven "point men" (the seven Mercury astronauts) for the American space program unified the nation behind the exploration of space. His 5 May 1961 suborbital Mercury mission established that the United States could send to and recover from space an individual. It was an enormously significant event for the United States.

Only recently had the nation been shocked by several outstanding space exploits from its closest rival, the Soviet Union—the 1957 orbiting of the *Sputnik* satellite, and Yuri Gagarin's space flight—and there was much pressure to regain national honor in the United States' own space program. The space flight made Shepard a national hero, but his strong personality and public countenance also underscored his stature among Americans as a role model.

Shepard's other great space flight took place on 31 January–9 February 1971, a decade later. (A medical disorder had kept him off flight status for several years.) Shepard commanded *Apollo 14* on a lunar landing mission at a significant time, a few months after the near-tragic *Apollo 13* mission in which the lunar lander had been used as a "lifeboat" for the crew. His mission, a complete success, boosted the national spirit and restored faith in the Apollo program. The achievements of *Apollo 14* were many: first use of the Mobile Equipment Transporter; placement of the largest payload ever in lunar orbit; longest stay on the lunar surface (33 hours); longest lunar surface EVA (9 hours and 17 minutes); first use of shortened lunar orbit rendezvous techniques; first use of color television on the lunar surface; first extensive orbital scientific experimentation period conducted in lunar orbit; and even the first lunar golf game (Shepard, an avid golfer, hit a hole in one).

Shepard left NASA in 1975 to pursue other business opportunities. In 1994 he published his life story, *Moonshot*, co-authored with fellow astronaut Deke Slayton.

Konstantin E. Tsiolkovskiy (1857–1935)

One of the earliest scientific theorists for the possibilities of space exploration was a Russian, Konstantin Eduardovich Tsiolkovskiy. Born on 17 September 1857 in the village of Izhevskoye, Spassk District, Ryazan Gubernia, he became enthralled with the possibilities of interplanetary travel as a boy. At age 14 he started independent study using books from his father's library on natural science and mathematics. He also developed a passion for invention and constructed balloons, propelled carriages, and other instruments.

To further his education, his parents sent young Tsiolkovskiy to Moscow to pursue technical studies. He stayed there only three years, however, returning home to become a tutor in mathematics and physics. In the process he completed his own education. In 1878 he passed the required examinations and received a diploma to pursue work as a "people's school teacher," a teacher in the Russian equivalent of an American high school or a German gymnasium. He obtained a teaching position in arithmetic and geometry at the district school in Borosck, Kaluga Province, north of Moscow. He remained in the Kaluga area for the rest of his career.

Tsiolkovskiy demonstrated genius in scientific matters. In 1881, for instance, he broke new ground with an article on the fundamentals of the kinetic theory of gases. His second publication, *The Mechanics of a Living Organism*, earned him election into the Society of Physics and Chemistry in St. Petersburg. Other publications, *The Problem of Flying by Means of Wings* (1890–1891) and *Elementary Studies of the Airship and Its Structure* (1898), showed Tsiolkovskiy's growing fascination with flight.

Tsiolkovskiy first started writing about space in 1898, when he submitted an article for publication to the Russian journal *Nauchnoye Obozreniye (Science Review)*. The article, "Investigating Space with Rocket Devices," presented years of calculations that laid out many of the principles of modern space flight and opened the door to future writings on the subject. Tsiolkovskiy described in depth the use of rockets for launching orbital space ships. This article finally was published in 1903.

There followed a series of increasingly sophisticated studies on the technical aspects of space flight. In the 1920s and 1930s Tsiolkovskiy was especially productive, publishing ten major works, clarifying the nature of bodies in orbit, developing scientific principles behind reaction vehicles, designing orbital space stations, and promoting interplanetary travel. He also expanded the scope of studies on many principles commonly used in rockets today: specific impulse to gauge engine performance, multistage boosters, fuel mixtures such as liquid hydrogen and liquid oxygen, the problems and possibilities inherent in microgravity, the promise of solar power, and spacesuits for extravehicular activity. Significantly, he never had the resources—perhaps not even the inclination—to experiment with rockets himself.

After the Bolshevik revolution of 1917 and the creation of the Soviet Union, Tsiolkovskiy was formally recognized for his accomplishments in the theory of space flight. Among other honors, in 1921 he received a lifetime pension from the state that allowed him to retire from teaching at age 64. Thereafter he devoted his full attention to developing spaceflight theories. He died at his home in Kaluga on 19 September 1935. His theoretical work greatly influenced later rocketeers, both in his native land and throughout Europe.

Although he was less well known during his lifetime in the United States, Tsiolkovskiy's work enjoyed broad study in the 1950s and 1960s as Americans sought to understand how the Soviet Union had accomplished such unexpected success in its early efforts in space flight. American space scientists then realized that his theoretical efforts had been essential for the development of the practical rocketry on which the Soviet space program was based.

James A. Van Allen (1914–)

James A. Van Allen was a pathbreaking astrophysicist best known for his work in magnetospheric physics. He was born on 7 September 1914 in Mount Pleasant, Iowa, where his father practiced law. From an early age Van Allen was intrigued by scientific theories and mechanical and electrical devices. He spent considerable time building electric motors, radios, and other mechanical equipment. He recalled that two highlights in his early life were building a Tesla coil to produce foot-long electrical discharges that made his hair stand on end and terrified his mother, and disassembling and reassembling the transmission of the family's Model T Ford.

In school Van Allen loved everything relating to science and mathematics, and he excelled in those areas as well as in wood shop and other classes where his mechanical skills were used. After completing high school in 1931, he attended Iowa Wesleyan College in Mount Pleasant. He enrolled in science and mathematics courses, once again standing out from his classmates in those areas. There he also became involved in geophysical research, studying the geomagnetic field of Earth by means of the theodolite on the magnetometer. His studies progressed systematically from there, and Van Allen became known as one of the brightest students at the college. When he graduated in 1935, he and most of the townspeople realized that his future lay in science—and not in Mount Pleasant.

He enrolled at the California Institute of Technology in the fall of 1935 and completed a M.S. the next year. From there he proceeded to the Ph.D., defending his dissertation in June 1939. Thereafter he accepted employment with the Department of Terrestrial Magnetism at the Carnegie Institution of Washington. Under the watchful eye of the legendary Merle Tuve, department director, Van Allen completed a Carnegie Research Fellowship on the measurement of the absolute cross-section for photodisintegration of the deuteron by 6.2 MeV gamma rays from protons to fluorine. He also became involved in several other research projects at the Carnegie Institution of Washington, some of them related to military research in anticipation of the U.S. entry into World War II.

In April 1942 Van Allen moved to the Applied Physics Laboratory at Johns Hopkins University, where he worked first to develop a rugged vacuum tube. He also helped develop proximity fuses for weapons used in the war, especially for torpedoes used by the U.S. Navy. By the fall of 1942 he had been commissioned as an officer in the Navy and was sent to the Pacific to field-test and complete operational requirements for the proximity fuses. Van Allen spent the remainder of the war either at the Applied Physics Laboratory or in battlefield environments undertaking real-time application of his equipment.

After completing his assignments in World War II, Van Allen returned to civilian life and began working in high altitude research, first for the Applied Physics Laboratory and then, after 1950, at the University of Iowa. Using captured V-2s and other rockets developed in the postwar era, Van Allen continued studying the magnetosphere of Earth in a series of increasingly complex upper-atmosphere experiments. These investigated the properties of cosmic rays, solar ultraviolet, high altitude photography, atmospheric ozone, and ionosphere current systems. By the mid-1950s, Van Allen had developed a reputation as a pathbreaking scientist in these increasingly important areas of research.

Van Allen's career took an important turn in 1955 when he and several other American scientists developed proposals for the launch of a scientific satellite as part of the research program conducted during the International Geophysical Year (IGY) of 1957–1958. Van Allen supported the proposal of the Army's Redstone Arsenal, offering a satellite to study terrestrial magnetism. But Project Vanguard, proposed by the Navy, was chosen over the Army's Explorer satellite on 9 September 1955.

After the success of the Soviet Union with *Sputnik 1*, Van Allen's Explorer spacecraft was approved for launch on a Redstone rocket. It flew on 31 January 1958 and returned highly important data about the radiation belts encircling Earth. Van Allen became a celebrity because of the success of this mission, and he pursued other important scientific projects in space thereafter. The radiation belts he discovered now bear his name, and the discipline of magnetospherics became important in part because of his initial work. In one way or another Van Allen was involved in the first four Explorer probes, the first Pioneers, several Mariner efforts, and the orbiting geophysical observatory.

James A. Van Allen retired from the University of Iowa in 1985 to become Carver Professor of Physics, Emeritus, after having served since 1951 as the head of the Department of Physics and Astronomy.

Wernher von Braun (1912–1977)

Wernher von Braun was one of the most important rocket developers and champions of space exploration during the period between the 1930s and the 1970s. The son of a German noble, Magnus Maximilian von Braun, the young spaceflight enthusiast was born in Wilintz, Germany, on 23 March 1912. As a youth he became fascinated by the possibilities of space exploration by reading the science fiction of Jules Verne and H. G. Wells, and from the scientific writings of Hermann Oberth, whose 1923 classic study, *Die Rakete zu den Planetenräumen (By Rocket to Space)*, prompted the young von Braun to master calculus and trigonometry so he could understand the physics of rocketry.

Even as a teenager von Braun had held a keen interest in space flight, becoming involved in the German rocket society, Verein für Raumschiffarht (VfR), as early as 1929. As a means of furthering his desire to build large and capable rockets, in 1932 he went to work for the German army to develop ballistic missiles. When Hitler came to power in 1933, von Braun remained in Germany and continued to work for the army.

While engaged in this work, on 27 July 1934 von Braun received a Ph.D. in aerospace engineering. Throughout the 1930s von Braun continued to develop rockets for the German army, and by 1941 designs had been developed for the ballistic missile that eventually became the V-2. The brainchild of Wernher von Braun's rocket team operating at a secret laboratory at Peenemünde on the Baltic coast, this rocket was the immediate antecedent of those used in space exploration programs in the United States and the Soviet Union. A liquid propellant missile extending some 46 feet in length and weighing 27,000 pounds, the V-2 flew at speeds in excess of 3,500 miles per hour and delivered a 2,200-pound warhead to a target 500 miles away. First flown in October 1942, it was used against targets in Europe beginning in September 1944. On 6 September, for instance, more than 6,000 German troops were deployed to Holland and northern Germany to bomb Belgium, France, and London with the newly developed V-2s.

Beginning on 8 September 1944, these forces began launching V-2s against Allied cities, especially Antwerp, Belgium, and London, England. By the end of the war 1,155 had been fired against England and another 1,675 had been launched against Antwerp and other continental targets. The guidance system for these missiles was imperfect, and many did not reach their targets; but they struck without warning and there was no defense against them. As a result, the V-2s had a terror factor far beyond their capabilities.

By the beginning of 1945 it was obvious to von Braun that Germany would not achieve victory against the Allies, and he began planning for the postwar era. Before the Allied capture of the V-2 rocket complex, von Braun arranged the surrender of 500 of his best rocket scientists, along with plans and test vehicles, to the Americans. For 15 years after World War II, von Braun worked with the U.S. Army in the development of ballistic missiles.

Because of the intriguing nature of V-2 technology, von Braun and his chief assistants achieved near-celebrity status inside the American military establishment. As part of a military operation called Project Paperclip, he and his "rocket team" were transported from defeated Germany to America, where they were installed at Fort Bliss, Texas. There they worked on rockets for the U.S. Army, launching them at White Sands Proving Ground, New Mexico. In 1950 von Braun's team moved to the Redstone Arsenal near Huntsville, Alabama, where they built the Army's Jupiter ballistic missile—and before that they built the Redstone, used by NASA to launch the first Mercury capsules. In 1960 von Braun's rocket development center was transferred from the Army to the newly established NASA and received a mandate to build the giant Saturn rocket. Accordingly, von Braun became director of NASA's Marshall Space Flight Center and the chief architect of the Saturn V launch vehicle, the superbooster that propelled Americans to the moon in the 1960s and early 1970s.

Von Braun also became one of the most prominent spokesmen of space exploration in the United States in the 1950s. In 1952 he gained note as a participant in an important symposium dedicated to the subject, and he burst on the nation's stage in the fall of 1952 with a series of articles in *Collier's*, a popular weekly periodical of the era. He also became a household name following his appearance on three Disney television shows dedicated to space exploration in the mid-1950s.

In 1970 the NASA leadership asked von Braun to move to Washington, D.C., to head the strategic planning effort for the agency. He left his home in Huntsville, Alabama, but in less than two years he decided to retire from NASA and work for Fairchild Industries of Germantown, Maryland. He died in Alexandria, Virginia, on 16 June 1977.

James E. Webb (1906–1992)

James Edwin Webb was the second administrator of the National Aeronautics and Space Administration (NASA), serving between 1961 and 1968. During his tenure, NASA developed the modern techniques necessary to coordinate and direct the most unique and complex technological enterprise in human history: the sending of human beings to the moon and bringing them safely back to Earth.

Born on 7 October 1906 in Granville County, North Carolina, he was the son of John Frederick Webb and Sarah Gorham. Webb was educated at the University of North Carolina, where he received an A.B. in education in 1928. He also studied law at George Washington University and was admitted to the Bar of the District of Columbia in 1936.

Webb enjoyed a long career in public service, coming to Washington, D.C., in 1932 and serving as secretary to Congressman Edward W. Pou, 4th North Carolina District, chairman of House Rules Committee, until 1934. He then served as assistant in the office of O. Max Gardner, attorney and former governor of South Carolina, in Washington, D.C., between 1934 and 1936. In 1936 Webb became secretary-treasurer and later vice president of the Sperry Gyroscope Company in Brooklyn, New York, before entering the U.S. Marine Corps in 1944. After World War II, Webb returned to Washington and served as executive assistant to O. Max Gardner, by then undersecretary of the Treasury, before being named director of the Bureau of the Budget in the Executive Office of the President, a position he held until 1949. President Harry S. Truman then asked Webb to serve as undersecretary of State. When the Truman administration ended early in 1953, Webb left Washington for a position at the Kerr-McGee Oil Corporation in Oklahoma. Webb's long experience in Washington was very useful during his years at NASA, where he lobbied for federal support for the space program and dealt with competing interests on Capitol Hill and in the White House.

James Webb returned to Washington on 14 February 1961 when he accepted the position of administrator of NASA. For seven years after President Kennedy's 1961 lunar landing announcement, through October 1968, James Webb politicked, coaxed, cajoled, and maneuvered for NASA in Washington. The longtime Washington insider was a master at bureaucratic politics. In the end, through a variety of methods, Administrator Webb built a seamless web of political liaisons that brought continued support for and resources to accomplish the Apollo moon landing on the schedule President Kennedy had announced. Webb left NASA in October 1968, just as Apollo was nearing a successful completion.

After retiring from NASA, Webb remained in Washington, D.C., and served on several advisory boards, including a stint as regent of the Smithsonian Institution. He died on 27 March 1992 in Washington, D.C.

The launch of the first American into space, Alan B. Shepard, Jr., on 5 May 1961 in the *Freedom 7* Mercury spacecraft, atop a converted ballistic missile, the Redstone, from the launch complex that eventually became the Kennedy Space Center, Cape Canaveral, Florida. (NASA photo number 61-MR3-72A)

In the mid-1960s NASA conducted its second human space flight project, Gemini, as a bridge between the early Mercury program and the Apollo effort to land Americans on the moon by the end of the decade. Here the *Gemini III* mission is being prepared for launch in March 1965. The two-astronaut Gemini spacecraft sits atop the *Titan II* launch vehicle at Launch Complex 19 at the Kennedy Space Center. (NASA photo number 65-H-374)

Some of the most realistic training for the weightlessness of space, albeit for limited periods, takes place in a KC-135 aircraft flying a parabolic flight path. During a dive weightlessness can be simulated for about 90 seconds. Here *Apollo 11* Lunar Module pilot Edwin E. Aldrin, Jr., practices working in space in his spacesuit. (NASA photo number 69-H-1097)

The launch of *Apollo 11* on 16 July 1969 for its historic flight to land on the moon. This photograph shows the press site for the launch, with more than 3,500 journalists on hand to cover the event. It was located 3.5 miles away from Launch Complex 39A. (NASA photo number 69-H-1134)

Astronauts Neil A. Armstrong and Edwin E. Aldrin, Jr., are aboard this Lunar Module as it leaves the *Apollo 11* Command Module to land on the moon on 20 July 1969. These astronauts spent 2 hours and 20 minutes on the lunar surface before returning to the Command Module and their trip back to Earth. (NASA photo number 69-H-1376)

Apollo 11 Mission Control at the Manned Spacecraft Center (renamed the Johnson Space Center in 1973) in Houston, Texas, broke into celebration at the time of the first lunar landing in 1969. The screen at the back of the photograph flashed the phrase, "Task Accomplished, July 1969," to show that the 1961 presidential commitment to land an American on the moon before the end of the decade had been completed. (NASA photo number 69-H-1299)

After the return of the *Apollo 11* astronauts to Earth, and their quarantine for several days to ensure they were not carrying alien viruses, the City of New York held a ticker-tape parade on 13 August 1969 to welcome Neil Armstrong, Buzz Aldrin, and Michael Collins back to Earth. The parade, traveling down Broadway and Park Avenue, was believed to be the largest in the city's history. (NASA photo number 69-H-1420)

One of the most striking images to emerge from the Apollo program, "Earthrise," was taken during the flight of *Apollo 8* in December 1968. Earth is about five degrees above the horizon with the moon directly below. This was the first time humanity saw its home from afar, a tiny, bright "blue marble" hanging in the blackness and the bleakness of space. (NASA photo number 68-H-1401)

This artist's conception shows the Pioneer spacecraft in orbit around the sun to study the interplanetary magnetic field, the solar wind, and cosmic rays. (NASA photo number 68-H-1102)

This photograph shows the Soviet Union's 1969 *Venera 5* spacecraft sent to Venus. (NASA photo number SL69-1067)

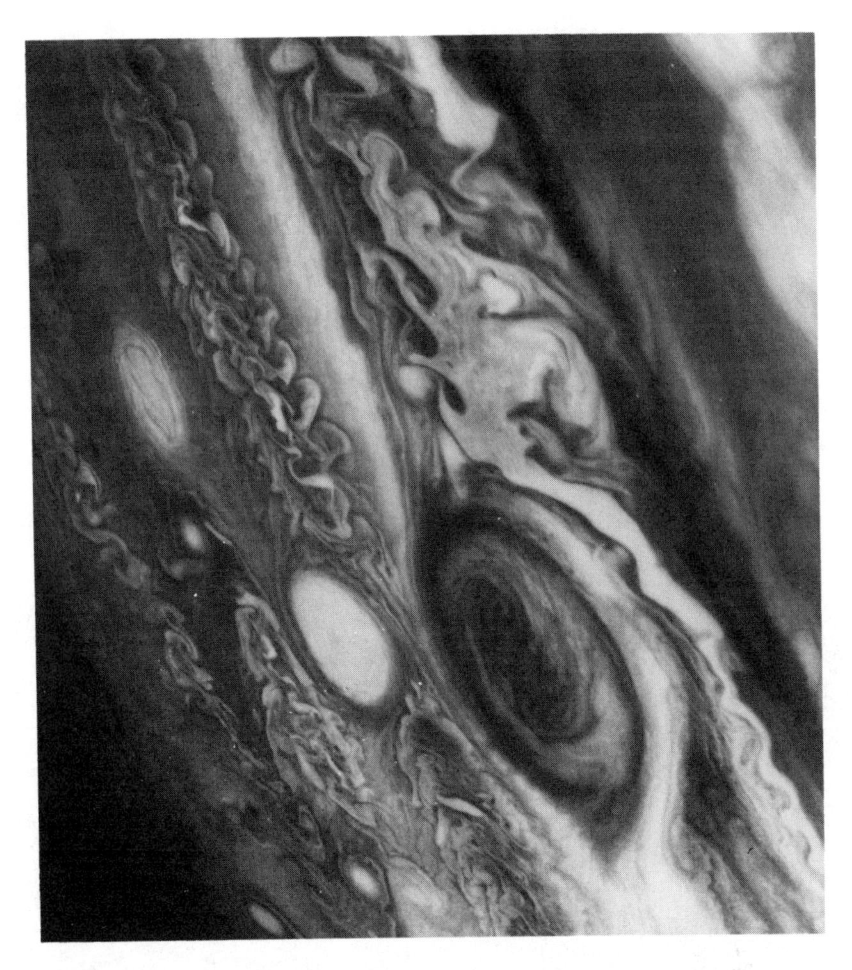

In the latter 1970s and early 1980s, the two Voyager probes to the outer planets of the Solar System provided stunning scientific data and visual images of those far distant worlds. Here, a *Voyager 2* image of Jupiter shows the "Great Red Spot," a catastrophic storm that has been raging in the planet's atmosphere since humans first glimpsed the details of Jupiter beginning in the seventeenth century. (NASA photo number 79-H-375)

The Hubble Space Telescope, launched in 1990, revolutionized our understanding of space science by providing the clearest images available of our universe. Initially hampered by a spherical aberration on the lens, this telescope was fully repaired in a shuttle mission in December 1993. Here astronaut Story Musgrave, anchored on the end of the shuttle's Remote Manipulator System, prepared to be elevated to the top of the telescope to install protective covers on the magnetometers. Astronaut Jeffrey Hoffman, at the bottom of the picture, is positioned to assist Musgrave. (NASA photo number 94-H-16)

The launch of the Space Shuttle *Challenger* on 6 April 1984. (NASA photo number 84-H-150)

The *Sojourner* rover and undeployed ramps onboard the *Mars Pathfinder* spacecraft can be seen in this image, by the Imager for *Mars Pathfinder* (IMP) on July 4, 1997 (Sol 1). The microrover *Sojourner* is latched to the petal, and has not yet been deployed. The ramps are a pair of deployable metal reels which will provide a track for the rover as it slowly rolls off the lander, over the spacecraft's deflated airbags, and onto the surface of Mars. (NASA photo number 97-H-590)

Primary Documents of Space Exploration

DEVELOPING A RATIONAL SPACE POLICY

When the Soviet Union launched *Sputnik I* into orbit on 4 October 1957, it sent shock waves through the United States, prompting a widespread questioning of American scientific and technical know-how. A far-reaching national debate resulted. A consistent topic in these discussions was the issue of national security, and leaders in the Eisenhower administration determined that civilian and military operations in space should be considered two sides of the same coin. An important space policy declaration was issued in June 1958 by the National Security Council; this document expanded the scope of proposed space activities and emphasized their national security dimensions in the context of Cold War rivalry with the Soviet Union.

At the same time, the Eisenhower administration decided that it should organize a new government agency that would lead the effort to catch up with the Soviets. In a legislative meeting with the president on 4 February 1958, the subject of a new space agency was raised. President Eisenhower agreed that such an agency would be appropriate. What ensued was the National Aeronautics and Space Act of 1958, creating the National Aeronautics and Space Administration (NASA). The act was signed into law on 29 July 1958.

Document 1
U.S. POLICY ON OUTER SPACE
(20 JUNE 1958)

Introductory Note

This statement of U.S. Policy on Outer Space is designated Preliminary because man's understanding of the full implications of outer space is only in its preliminary stages. As man develops a fuller understanding of the new dimension of outer space, it is probable that the long-term results of exploration and exploitation will basically affect international and national political and social institutions.

Perhaps the starkest facts which confront the United States in the immediate and foreseeable future are (1) the USSR has surpassed the United States and the Free World in scientific and technological accomplishments in outer space, which have captured the imagination and admiration of the world; (2) the USSR, if it maintains its present superiority in the exploitation of outer space, will be able to use that superiority as a means of undermining the prestige and leadership of the United States; and (3) the USSR, if it should be the first to achieve a significantly superior military capability in outer space, could create an imbalance of power in favor of the Sino-Soviet Bloc and pose a direct military threat to U.S. security.

The security of the United States requires that we meet these challenges with resourcefulness and vigor.

General Considerations

Introduction

Significance of Outer Space to U.S. Security

1. More than by any other imaginative concept, the mind of man is aroused by the thought of exploring the mysteries of outer space.

2. Through such exploration, man hopes to broaden his horizons, add to his knowledge, improve his way of living on earth. Already, man is sure that through further exploration he can obtain certain scientific and military values. It is reasonable for man to believe that there must be, beyond these areas, different and great values still to be discovered.

3. The technical ability to explore outer space has deep psychological implications over and above the stimulation provided by the opportunity to explore the unknown. With its hint of the possibility of the discovery of fundamental truths concerning man, the earth, the solar system, and the universe, space exploration has an appeal to deep insights within man which transcend his earthbound concerns. The manner in which outer space is explored and the uses to which it is put thus take on an unusual and peculiar significance.

4. The beginning stages of man's conquest of space have been focused on technology and have been characterized by national competition. The result has been a tendency to equate achievement in outer space with leadership in science, military capability, industrial technology, and with leadership in general.

5. The initial and subsequent successes by the USSR in launching large earth satellites have profoundly affected the belief of peoples, both in the United States and abroad, in the superiority of U.S. leadership in science and military capability. This psychological reaction of sophisticated and unsophisticated peoples everywhere affects U.S. relations with its allies, with the Communist Bloc, and with neutral and uncommitted nations.

6. In this situation of national competition and initial successes by the USSR, further demonstrations by the USSR of continuing leadership in outer space capabilities might, in the absence of comparable U.S. achievements in this field, dangerously impair the confidence of these peoples in U.S. over-all leadership. To be strong and bold in space technology will enhance the prestige of the United States among the peoples of the world and create added confidence in U.S. scientific, technological, industrial and military strength.

7. The novel nature of space exploitation offers opportunities for international cooperation in its peaceful aspects. It is likely that certain nations may be willing to enter into cooperative arrangements with the United States. The willingness of the Soviets to cooperate remains to be determined. The fact that the results of cooperation in certain fields, even though entered into for peaceful purposes, could have military application, may condition the extent of such cooperation in those fields. . . .

Manned Exploration of Outer Space

24. In addition to satisfying man's urge to explore new regions, manned exploration of outer space is of importance to our national security because:

(a) Although present studies in outer space can be carried on satisfactorily by using only unmanned vehicles, the time will undoubtedly come when man's judgment and resourcefulness will be required fully to exploit the potentialities of outer space.

(b) To the layman, manned exploration will represent the true conquest of outer space. No unmanned experiment can substitute for manned exploration in its psychological effect on the peoples of the world.

(c) Discovery and exploration may be required to establish a foundation for the rejection of USSR claims to exclusive sovereignty of other planets which may be visited by nationals of the USSR.

25. The first step in manned outer space travel could be undertaken using rockets and components now under study and development. Travel by man to the moon and beyond will probably require the development of new basic vehicles and equipment. . . .

Comparison of USSR and U.S. Capabilities in Outer Space Activities

40. Conclusive evidence shows that the Soviets are conducting a well-planned outer space program at high priority. The table below attempts to estimate the U.S. and USSR timetable for accomplishment of specific outer space flight activities.

a. Soviet space flight capabilities estimated in the table reflect the earliest possible time periods in which each specific event could be successfully accomplished.

1. The space flight program is in competition with many other programs, particularly the missile program. The USSR probably cannot successfully accomplish all of the estimated space flight activities within the time periods specified. The USSR will not permit its space flight program to interfere with achieving an early operational capability for ICBMs (which enjoy the highest priority).
2. The USSR is believed to have the intention to pursue both an active space flight program designed to put man into outer space for military and/or scientific purposes, and further scientific research utilizing earth satellites, lunar rockets, and probes of Mars and Venus; but it cannot be determined, at this time, whether the basic scientific program or the "man in space" program enjoys the higher priority and will, therefore, be pursued first.

b. U.S. space flight capabilities indicated in the table reflect the earliest possible time periods in which each specific event could be successfully accomplished. Not all of the indicated activities could be successfully accomplished within the time period specified. It must also be recognized that the accomplishment of some of the activities listed would impinge upon space activities already programmed, or upon other military programs. . . .

Psychological Exploitation

51. In the near future, while the USSR has a superior capability in space technology, judiciously select . . . projects for implementation which, while having scientific or military value, are designed to achieve a favorable world-wide psychological impact.

52. Identify, to the greatest extent possible, the interests and aspirations of other Free World nations in outer space with U.S.-sponsored activities and accomplishments.

53. Develop information and other programs that will exploit fully U.S. outer space activities on a continuing basis; especially, during the period while the USSR has superior over-all outer space capabilities, those designed to counter the psychological impact of Soviet outer space activities and to present U.S. outer space progress in the most favorable comparative light.

Reconnaissance Satellites

54. In anticipation of the availability of reconnaissance satellites, seek urgently a political framework which will place the uses of U.S. reconnaissance satellites in a political and psychological context most favorable to the United States.

55. At the earliest technologically practicable date, use reconnaissance satellites to enhance to the maximum extent the U.S. intelligence effort.

International Cooperation in Outer Space Activities

56. . . . as a means of maintaining the U.S. position as the leading advocate of the use of outer space for peaceful purposes, be prepared to propose that the United States join with other nations, including the USSR, in cooperative efforts relating to outer space. . . .

Limited International Arrangements to Regulate Outer Space Activities

57. Propose international agreements concerning appropriate means for maintaining a full and current public record of satellite orbits and emission frequencies. . . .

Administration of Outer Space Programs

62. Provide through appropriate legislation for the conduct of U.S. outer space activities under the direction of a civilian agency, except in so far as such activities may be peculiar to or primarily associated with weapons systems or military operations, in the case of which activities the Department of Defense shall be responsible.

Annex A: The Soviet Space Program

1. Objectives and Scope of Program. Conclusive evidence shows that the Soviets are conducting a well-planned space flight program at high priority. This program is apparently aimed at placing both instrumented and manned vehicles into space. Certain successes have been exhibited already in the instrumented vehicles category (including the orbiting of three earth satellites, one containing a dog) and we believe they are fully capable of achieving manned space flight within the next few years.

2. General. Evidence of Soviet interest in space flight dates back to a publication in 1903 of a paper, "Investigation of Universal Space by Means of Rocket Flight," by the eminent Russian scientist Tsiolkovsky. This highly scientific treatise for the first time mathematically established the fundamentals of rocket dynamics and included a proposal for an artificial earth satellite. Reactive motion (rockets) was seriously engaged again in the latter '20s and in the '30s. In April 1955, the Interagency Commission for Interplanetary Communications was formed under the Academy of

Sciences to establish an automatic laboratory for scientific research in cosmic space as a first step in solving the problems of interplanetary travel. Since early 1955 several hundred articles on space research, earth satellites and space flight have been published in the USSR. Many of the articles have been written by high-caliber Soviet scientists and most deal with the theoretical principles of space flight.

3. Capabilities. The Soviet Union dramatically demonstrated its interest and current capability in space flight with the launching of two earth satellites in October and November 1957, and a third in mid-May 1958. The complex facilities and skills needed to operate the large rocket vehicles required for the launching of a satellite or space vehicle are apparently available within Soviet military. Thus, although the first space flights were doubtless undertaken for the furtherance of scientific knowledge and for whatever psychological and political advantage would accrue, the Soviet military department, by intimate participation of its hardware and personnel, is in a position to utilize immediately such knowledge for the enhancement of the Soviet military position and objectives. . . . We believe the depth and advancement of their research and development makes them world leaders in these areas. In particular their work in space medicine and cosmic biology are strong indicators of their serious intent to put man in space at an early date.

4. Time Scales.

a. The following milestones are considered at least partially affiliated with a space program and indicate historically the long-term interest of Soviet Union in this endeavor:

1903	Initial treatise on space flight
1923	Soviet Institute on Theoretical Astronomy founded
1929	First significant rocket studies conducted, "Group for the Investigation of Reactive Motion" founded
1934	Government-sponsored rocket research program established
1940	Flight of first Soviet rocket-powered aircraft
1946–47	Rocket-propelled intercontinental bomber program organized
1953–55	Systematic investigation of moon flight problems undertaken
1955	Interagency Commission for Interplanetary Communications established (Apr.)
1955–58	Over 500 Soviet articles published dealing with space research, earth satellites and manned space flight
1957	First artificial earth satellites orbited (Oct.–Nov.)

b. Future Capabilities. Soviet space flight capabilities estimated in this section are the earliest possible time periods in which each specific event could be successfully accomplished. It is recognized that the space flight program is in competition with many other programs, particularly the missile program, and that the USSR probably cannot successfully accomplish all of the estimated space flight activities within the time periods specified. We believe the USSR has the intention to pursue an active flight program designed to put man into space for military and/or scientific purposes. We also believe they have a definite intention to pursue further scientific research utilizing earth satellites, lunar rockets, and probes of Mars and Venus. We cannot, at this time, determine whether the basic scientific program or the "man in space" program enjoys the higher priority and will, therefore, be pursued first. Whichever approach is adopted will probably result in some slippage in the capability dates indicated for the other program. We believe the Soviet ICBM program still enjoys the highest priority and that the USSR will not permit its space flight program to interfere with achieving an early operational ICBM capability. . . .

(2) Lunar Rockets. The USSR has had the capability of launching a lunar probe toward the vicinity of the moon since the fall of 1957 as far as propulsion and guidance requirements are concerned. A Soviet program of lunar probes could commence with experimental rockets followed by rocket landings on the moon with increasingly heavy loads containing scientific and telemetering equipment. Placing a satellite into orbit around the moon requires the use of a retro-rocket and more accurate guidance. It is believed that the USSR could achieve a lunar satellite in late 1958–1959 and have a lunar soft landing about six months thereafter.

(3) Manned Earth Satellites. Sufficient scientific data could probably have been attained and recovery techniques perfected to permit the USSR to launch a manned satellite into orbital flight and recovery by about 1959–1960. A manned capsule-type satellite as well as a manned glide-type vehicle appear to be feasible techniques and within Soviet capabilities. However, it is believed that the first Soviet orbital recovery attempt will probably be with the manned capsule.

(4) Planetary Probes. Planetary probe vehicles could utilize existing Soviet ICBM propulsion units for the first stage and presently available guidance components. It is believed that the USSR could launch probes towards Mars and Venus with a good chance of success. The first launchings toward Mars could occur in August 1958, when Mars will be in the most favorable position relative to the earth. More sophisticated probes could occur in October 1960, when Mars will again be in a favorable position relative to the earth. Probes toward Venus could probably occur in June 1959, and more sophisticated probe vehicles could be launched in January 1961.

(5) Manned Circumlunar Flights. Contingent upon their success with

manned earth satellites and the development of a new, large booster engine, and concurrent advances in scientific experimentations with lunar rockets, the USSR could achieve a capability for manned circumlunar flight with reasonable chance for success in about 1961–1962.

(6) Manned Lunar Landings. It is not believed that the USSR will have a capability for manned lunar landings until some time after 1965.

(7) Space Platforms. There is insufficient information on the problems as well as the utility of constructing a platform in space to determine the Soviet capability. It is believed, however, that they are capable of placing a very large satellite (about 25,000 pounds) into orbit in 1961–1962 and that this vehicle could serve some of the scientific functions of a large space platform without the difficulties of joining and constructing such a platform in space. . . .

Source: National Security Council, NSC 5814, "U.S. Policy on Outer Space," 20 June 1958, Presidential Papers, Dwight D. Eisenhower Library, Abilene, KS.

Document 2
LEGISLATIVE MEETING, SUPPLEMENTARY NOTES (4 FEBRUARY 1958)

Outer Space Program—A question was raised as to whether a new Space Agency should be set up within Department of Defense (as provided in the pending Defense appropriation bill), or be set up as an independent agency. The President's feeling was essentially a desire to avoid duplication, and priority for the present would seem to rest with Defense because of paramountcy of defense aspects. However, the President thought that in regard to non-military aspects, Defense could be the operational agent, taking orders from some non-military scientific group. The National Science Foundation, for instance, should not be restricted in any way in its peaceful research.

Dr. Killian had some reservations as to the relative interest and activity of military vs. peaceful aspects, as did the Vice President, who thought our posture before the world would be better if non-military research in outer space were carried forward by an agency entirely separate from the military.

There was some discussion of the prospect of a lunar probe. Dr. Killian thought this might be next on the list of Russian efforts. He had some doubt as to whether the United States should at this late date attempt to press a lunar probe, but the question would be fully canvassed by the Science Advisory Committee in the broad survey it had under way. Dr. Killian thought the United States might do a lunar probe in 1960, or perhaps get to it on a crash program by 1959. Sen. Saltonstall had heard, however, that it might even be accomplished in 1958, if pressed hard enough.

Dr. Killian outlined for the Leadership the various phases of future development (along the lines of the subsequent press release listing projects in the "soon," "later," and "much later" categories).

Sen. Knowland complained about having to get his information about Space research from the Democratic Senator from Washington (Jackson)—which was just as bad as having to learn from Mr. Symington anything there was to know about the Air Force.

The President was firmly of the opinion that a rule of reason had to be applied to these Space projects—that we couldn't pour unlimited funds into these costly projects where there was nothing of early value to the Nation's security. He recalled the great effort he had made for the Atomic Peace Ship but Congress would not authorize it, even though in his opinion it would have been a very worthwhile project.

And in the present situation, the President mused, he would rather have a good Redstone than be able to hit the moon, for we didn't have any enemies on the moon.

Sen. Knowland pressed the question of hurrying along with a lunar probe, because of the psychological factor. He recalled the great impact of Sputnik, which seemed to negate the impact of our large mutual security program. If we are close enough to doing a probe, he said, we should press it. The President thought it might be OK to go ahead with it if it could be accomplished with some missile already developed or nearly ready, but he didn't want to just rush into an all-out effort on each one of these possible glamor performances without a full appreciation of their great cost. Also, there would have to be a clear determination of what agency would have the responsibility.

The Vice President reverted to the idea of setting up a separate agency for "peaceful" research projects, for the military would be deterred from things that had no military value in sight. The President thought Defense would inevitably be involved since it presently had all the hardware, and he did not want further duplication. He did not preclude having eventually a great Department of Space.

Source: L. Arthur Minnich Jr., "Legislative Meeting, Supplementary Notes," 4 February 1958, Presidential Papers, Dwight D. Eisenhower Library, Abilene, KS.

THE DECISION TO GO TO THE MOON

The documents that follow lay out the decision-making process that culminated in President John F. Kennedy's announcement on 25 May 1961 to commit the nation to landing an American on the moon by the end of the decade—possibly the most significant decision in the history of U.S. space exploration. In the 20 April 1961 memorandum reprinted here, Kennedy set the tone of the policy debate. Of all those who were consulted during this review, no one had been thinking longer about the subject than Wernher von Braun. In late April 1961, von Braun sent a letter privately to Vice President Lyndon Johnson that was important in sealing the direction of the decision-making process. He stated that the United States had "an excellent chance" of beating the Russians to a lunar landing.

On 25 May 1961, President Kennedy unveiled the commitment to execute Project Apollo before Congress in a speech on "Urgent National Needs," also excerpted here. Congress agreed to Kennedy's request with barely any comment. As seen in this excerpt from the speech, Kennedy discussed the space program in the context of the Cold War rivalry with the Soviet Union.

Document 3
MEMORANDUM FOR VICE PRESIDENT (20 APRIL 1961)

In accordance with our conversation I would like for you as Chairman of the the Space Council to be in charge of making an overall survey of where we stand in space.

1. Do we have a chance of beating the Soviets by putting a laboratory in space, or by a trip round the moon, or by a rocket to land on the moon, or by a rocket to go to the moon and back with a man? Is there any other space program which promises dramatic results in which we could win?

2. How much additional would it cost?

3. Are we working 24 hours a day on existing programs? If not, why not? If not, will you make recommendations to me as to how work can be speeded up?

4. In building large boosters should we put our emphasis on nuclear, chemical or liquid fuel, or a combination of these three?

5. Are we making maximum effort? Are we achieving necessary results?

I have asked Jim Webb, Dr. Weisner, Secretary McNamara and other responsible officials to cooperate with you fully. I would appreciate a report on this at the earliest possible moment.

[signed]
John F. Kennedy

Source: John F. Kennedy, "Memorandum for the Vice President," 20 April 1961, Presidential Papers, John F. Kennedy Presidential Library, Boston, MA.

Document 4
URGENT NATIONAL NEEDS (25 MAY 1961)

. . . Finally, if we are to win the battle that is now going on around the world between freedom and tyranny, the dramatic achievements in space which occurred in recent weeks should have made clear to us all, as did the Sputnik in 1957, the impact of this adventure on the minds of men everywhere, who are attempting to make a determination of which road they should take. Since early in my term, our efforts in space have been under review. With the advice of the Vice President, who is Chairman of the National Space Council, we have examined where we are strong and where we are not, where we may succeed and where we may not. Now it is time to take longer strides—time for this nation to take a clearly leading role in space achievement, which in many ways may hold the key to our future on earth.

I believe we possess all the resources and talents necessary. But the facts of the matter are that we have never made the national decisions or marshalled the national resources required for such leadership. We have never specified long-range goals on an urgent time schedule, or managed our resources and our time so as to insure their fulfillment.

Recognizing the head start obtained by the Soviets with their large rocket engines, which gives them many months of leadtime, and recognizing the likelihood that they will exploit this lead for some time to come in still more impressive successes, we nevertheless are required to make new efforts on our own. For while we cannot guarantee that we shall one day be first, we can guarantee that any failure to make this effort will make us last. We take an additional risk by making it in full view of the world, but as shown by the feat of astronaut Shepard, this very risk enhances our stature when we are successful. But this is not merely a race. Space is open to us now; and our eagerness to share its meaning is not governed by the efforts of others. We go into space because whatever mankind must undertake, free men must fully share.

I therefore ask the Congress, above and beyond the increases I have earlier requested for space activities, to provide the funds which are needed to meet the following national goals:

First, I believe that this nation should commit itself to achieving the goal, before this decade is out, of landing a man on the moon and returning him safely to the earth. No single space project in this period will be more impressive to mankind, or more important for the long-range exploration of space; and none will be so difficult or expensive to accomplish. We propose to accelerate the development of the appropriate lunar space craft. We propose to develop alternate liquid and solid fuel boosters, much larger than any now being developed, until certain which is superior. We propose additional funds for other engine development and for unmanned explorations—explorations which are particularly important for one purpose which this nation will never overlook; the survival of the man who first makes this daring flight. But in a very real sense, it will not be one man going to the moon—if we make this judgment affirmatively, it will be an entire nation. For all of us must work to put him there. . . .

Let it be clear—and this is a judgment which the Members of the Congress must finally make—let it be clear that I am asking the Congress and the country to accept a firm commitment to a new course of action—a course which will last for many years and carry very heavy costs: 531 million dollars in fiscal '62—and estimated seven to nine billion dollars additional over the next five years. If we are to go only half way, or reduce our sights in the face of difficulty, in my judgment it would be better not to go at all.

Now this is a choice which this country must make, and I am confident that under the leadership of the Space Committees of the Congress, and the Appropriating Committees, that you will consider the matter carefully.

It is a most important decision that we make as a nation. But all of you have lived through the last four years and have seen the significance of space and the adventure in space, and no one can predict with certainty what the ultimate meaning will be of mastery of space.

I believe we should go to the moon. But I think every citizen of this country as well as the Members of the Congress should consider the matter carefully in making their judgment, to which we have given attention over many weeks and months, because it is a heavy burden, and there is no sense in agreeing or desiring that the United States take an affirmative position in outer space, unless we are prepared to do the work and bear the burdens to make it successful. If we are not, we should decide today and this year.

This decision demands a major national commitment of scientific and technical manpower, material and facilities, and the possibility of their diversion from other important activities where they are already thinly spread. It means a degree of dedication, organization and discipline which have not always characterized our research and development efforts. It means we cannot afford undue work stoppages, inflated costs of material or talent, wasteful interagency rivalries, or a high turnover of key personnel.

New objectives and new money cannot solve these problems. They could, in fact, aggravate them further—unless every scientist, every engineer, every serviceman, every technician, contractor, and civil servant gives his personal pledge that this nation of freedom aggravate them further in the exciting adventure of space . . .

Source: Speech on "Urgent National Needs," 25 May 1961, *Public Papers of the Presidents of the United States: John F. Kennedy, 1961* (Washington, DC: Government Printing Office, 1962), pp. 403–405.

THE POSSIBILITY OF AMERICAN/SOVIET COOPERATION IN SPACE

Although Project Apollo was a tangible expression of Cold War rivalry between the United States and the Soviet Union, recently declassified documents have demonstrated that there were efforts on the part of the Kennedy administration in the early 1960s to make it a cooperative program. President Kennedy had called for U.S.-Soviet space cooperation in his 20 January 1961 inaugural address and in his first State of the Union Address a few days later. To examine the possibilities for such cooperation, the presidential science adviser, Jerome Weisner, set up both an external advisory group and an internal government study group. However, as this effort was nearing completion the Soviet Union on 12 April 1961 launched Yuri Gagarin into orbit. A few days later President Kennedy decided that the United States had to compete, not cooperate, in major space activities such as Apollo, and Weisner's study was temporarily set aside.

Yet Kennedy still hoped that other areas of space could serve as arenas for cooperation. When Nikita Khrushchev on 21 February 1962 congratulated the United States on its first human orbital flight, the *Freedom* 7 Mercury mission of John Glenn, it seemed to open the door to such cooperation. Kennedy replied immediately in the memorandum printed here, proposing specific cooperative initiatives with the Soviet Union. This move marked the beginning of substantive cooperation between the two space superpowers, but not in human space flight.

Eighteen months later, when President Kennedy met with NASA administrator James E. Webb on 15 September 1963, the political climate seemed right for a substantial cooperative project with the Soviet Union—in part because of the high levels of spending for Apollo. Kennedy's national security adviser McGeorge Bundy, suggested to the president that a cooperative mission would be desirable if it was technically, institutionally, and politically feasible. Two days later, in an address to the United Nations General Assembly, Kennedy suggested that the United States and the Soviet Union take the lead in making the first human voyages to the moon an undertaking of all countries.

Ten days before he was assassinated, President Kennedy signed a final memorandum giving the NASA administrator the lead within the Executive Branch in developing substantive proposals for enhanced U.S.-U.S.S.R. space cooperation—including a lunar landing—as a direct follow-up to his speech at the United Nations. Following the president's assassination in Dallas on 22 November 1963, Apollo became viewed as a fitting tribute to the fallen leader and could not thereafter be changed appreciably. It remained an entirely U.S. project until its conclusion in December 1972. Had President Kennedy lived, the American effort to land on the moon might have unfolded in an entirely different manner.

Document 5
PRESIDENT KENNEDY'S LETTER TO NIKITA KHRUSHCHEV (7 MARCH 1962)

Dear Mr. Chairman:

On February twenty-second last I wrote you that I was instructing appropriate officers of this Government to prepare concrete proposals for immediate projects of common action in the exploration of space. I now present such proposals to you.

The exploration of space is a broad and varied activity and the possibilities for cooperation are many. In suggesting the possible first steps which are set out below, I do not intend to limit our mutual consideration of desirable cooperative activities. On the contrary, I will welcome your concrete suggestions along these or other lines.

1. Perhaps we could render no greater service to mankind through our space programs than by the joint establishment of an early operational weather satellite system. Such a system would be designed to provide global weather data for prompt use by any nation. To initiate this service, I propose that the United States and the Soviet Union each launch a satellite to photograph cloud cover and provide other agreed meteorological services for all nations. The two satellites would be placed in near-polar orbits in planes approximately perpendicular to each other, thus providing regular coverage of all areas. This immensely valuable data would then be disseminated through normal international meteorological channels and would make a significant contribution to the research and service programs now under study by the World Meteorological Organization in response to Resolution 1721 (XVI) adopted by the United Nations General Assembly on December 20, 1961.

2. It would be of great interest to those responsible for the conduct of our respective space programs if they could obtain operational tracking services from each other's territories. Accordingly, I propose that each of our countries establish and operate a radio tracking station to provide tracking services to the other, utilizing equipment which we would each provide to the other. Thus, the United States would provide the technical equipment for a tracking station to be established in the Soviet Union and to be operated by Soviet technicians. The United States would in turn establish and operate a radio tracking station utilizing Soviet equipment. Each country would train the other's technicians in the operation of its equipment, would utilize the station located on its territory to provide tracking services to the other, and would afford such access as may be necessary to accommodate modification and maintenance of equipment from time to time.

3. In the field of the earth sciences, the precise character of the earth's magnetic field is central to many scientific problems. I propose therefore that we cooperate in mapping the earth's magnetic field in space by utilizing two satellites, one in a near-earth orbit and the second in a more distant orbit. The United States would launch one of these satellites while the Soviet Union would launch the other. The data would be exchanged throughout the world scientific community, and opportunity for correlation of supporting data obtained on the ground would be arranged.

4. In the field of experimental communications by satellite, the United

States has already undertaken arrangements to test and demonstrate the feasibility of intercontinental transmissions. A number of countries are constructing equipment suitable for participation in such testing. I would welcome the Soviet Union's joining in this cooperative effort which will be a step toward meeting the objective, contained in United Nations General Assembly Resolution 1721 (XVI), that communications by means of satellites should be available to the nations of the world as soon as practicable on a global and non-discriminatory basis. I note also that Secretary Rusk has broached the subject of cooperation in this field with Minister Gromyko and that Mr. Gromyko has expressed some interest. Our technical representatives might now discuss specific possibilities in this field.

5. Given our common interest in manned space flights and in insuring man's ability to survive in space and return safely, I propose that we pool our efforts and exchange our knowledge in the field of space medicine, where future research can be pursued in cooperation with scientists from various countries.

Beyond these specific projects we are prepared now to discuss broader cooperation in the still more challenging projects which must be undertaken in the exploration of outer space. The tasks are so challenging, the costs so great, and the risk to the brave men who engage in space exploration so grave, that we must in all good conscience try every possibility of sharing these tasks and costs and of minimizing these risks. Leaders of the United States space program have developed detailed plans for an orderly sequence of manned and unmanned flights for exploration of space and the planets. Out of discussion of these plans, and of our own, for undertaking the tasks of this decade would undoubtedly emerge possibilities for substantive scientific and technical cooperation in manned and unmanned space investigation. Some possibilities are not yet precisely identifiable, but should become clear as the space programs of our two countries proceed.

In the case of others it may be possible to start planning together now. For example, we might cooperate in unmanned exploration of the lunar surface, or we might commence now the mutual definition of steps to be taken in sequence for an exhaustive scientific investigation of the planets Mars or Venus, including consideration of the possible utility of manned flight in such programs. When a proper sequence for experiments has been determined, we might share responsibility for the necessary projects. All data would be made freely available.

I believe it is both appropriate and desirable that we take full cognizance of the scientific and other contributions which other states the world over might be able to make in such programs. As agreements are reached between us on any parts of these or similar programs, I propose that we report them to the United Nations Committee on the Peaceful Uses of Outer Space. The Committee offers a variety of additional opportunities for joint cooperative efforts within the framework of its mandate as set forth in General Assembly Resolutions 1472 (XIV) and 1721 (XVI).

I am designating technical representatives who will be prepared to meet and discuss with your representatives our ideas and yours in a spirit of practical cooperation. In order to accomplish this at an early date I suggest that the representatives of our countries, who will be coming to New York to take part in the United Nations Outer Space Committee, meet privately to discuss the proposals set forth in this letter.

Sincerely,
[signed]
John F. Kennedy

Source: John F. Kennedy, "Text of Letter Dated March 7, 1962, From President Kennedy to Chairman Khruschev, Re: Cooperation in Peaceful Uses of Outer Space," 7 March 1962, NASA Historical Reference Collection, History Office, NASA Headquarters, Washington, DC.

Document 6
MEMORANDUM FOR THE PRESIDENT (18 SEPTEMBER 1963)

SUBJECT: *Your 11 A.M. appointment with Jim Webb*

Webb called me yesterday to comment on three interconnected aspects of the space problem that he thinks may be of importance in his talk with you:

1. *Money*. The space authorization is passed at $5,350 billion, and he expects the appropriation to come out at about $5,150 billion. While the estimates are not complete, his current guess is that in early '64 he will require a supplemental of $400 million ($200 million requiring authorization and $200 million appropriation only) in order to keep our commitment to a lunar landing in the 1960's.

2. *The Soviets*. He reports more forthcoming noises about cooperation from Blagonravov in the UN, and I am trying to run down a report in today's *Times* . . . that we have rebuffed the Soviets on this. Webb himself is quite open to an exploration of possible cooperation with the Soviets and thinks that they might wish to use our big rocket, and offer in exchange the advanced technology which they are likely to get in the immediate future. (For example, Webb expects a Soviet landing of instruments on the moon to establish moon-earth communications almost any time.)

The obvious choice is whether to press for cooperation or to continue to use the Soviet space effort as a spur to our own. The *Times* story suggests that there is already low-level disagreement on exactly this point.

3. *The Military Role*. Webb reports that the discontent of the military

with their limited role in space damaged the bill on the Hill this year, with no corresponding advantage to the military. He thinks this point can and should be made to the Air Force, and he believes that the thing to do is to offer the military an increased role somehow. He has already had private exploratory talks with Ros Gilpatric for this purpose.

Webb thinks the best place for a military effort in space would be in the design and manning of a space craft in which gravity could be simulated, in preparation for later explorations. He thinks such a space craft may be the next logical step after Gemini. On the other hand, he is quite cool about the use of Titan III and Dyna-soar and would be glad to see them both canceled. You will recall that McNamara has just come out on the other side of Titan III.

My own hasty judgment is that the central question here is whether to compete or to cooperate with the Soviets in a manned lunar landing:

1. *If we compete*, we should do everything we can to unify all agencies of the United States Government in a combined space program which comes as near to our existing pledges as possible.

2. *If we cooperate*, the pressure comes off, and we can easily argue that it was our crash effort of '61 and '62 which made the Soviets ready to cooperate.

I am for cooperation if it is possible, and I think we need to make a really major effort inside and outside the government to find out whether in fact it can be done. Conceivably this is a better job for Harriman than East-West trade, which might almost as well be given to George Ball.

[initialed by McGeorge Bundy]
McG.B

Source: McGeorge Bundy, Memorandum for the President, "Your 11 A.M. appointment with Jim Webb," 18 September 1963, NASA Historical Reference Collection, History Office, NASA Headquarters, Washington, DC.

Document 7
NATIONAL SECURITY ACTION MEMORANDUM NO. 271 (12 NOVEMBER 1963)

MEMORANDUM FOR The Administrator, National Aeronautics and Space Administration

SUBJECT: Cooperation with the USSR on Outer Space Matters

I would like you to assume personally the initiative and central responsibility within the Government for the development of a program of substantive cooperation with the Soviet Union in the field of outer space, including the development of specific technical proposals. I assume that you will work closely with the Department of State and other agencies as appropriate.

These proposals should be developed with a view to their possible discussion with the Soviet Union as a direct outcome of my September 20 proposal for broader cooperation between the United States and the USSR in outer space, including cooperation in lunar landing programs. All proposals or suggestions originating within the Government relating to this general subject will be referred to you for your consideration and evaluation.

In addition to developing substantive proposals, I expect that you will assist the Secretary of State in exploring problems of procedure and timing connected with holding discussions with the Soviet Union and in proposing for my consideration the channels which would be most desirable from our point of view. In this connection the channel of contact developed by Dr. [Hugh L.] Dryden between NASA and the Soviet Academy of Sciences has been quite effective, and I believe that we should continue to utilize it as appropriate as a means of continuing the dialogue between the scientists of both countries.

I would like an interim report on the progress of our planning by December 15.

[signed]
John F. Kennedy

Source: National Security Action Memorandum No. 271, "Cooperation with the USSR on Outer Space Matters," 12 November 1963, NASA Historical Reference Collection, History Office, NASA Headquarters, Washington, DC.

THE GEMINI INTERLUDE

In the history of human space flight, most Americans remember the first hesitant flights of the Mercury program and its legendary astronaut corps led by Alan Shepard and John Glenn. Many also remember the actual lunar landings of Project Apollo. Few, however, recall the critical knowledge necessary to accomplish Apollo provided by the second NASA space flight program of the 1960s, Project Gemini. The two-person Gemini spacecraft provided important data on maneuvering in space (both in vehicles and during extravehicular activities), rendezvous and docking of spacecraft in orbit, and capsule guidance and control. This recollection by Robert R. Gilruth, director of the Manned Spacecraft Center in Houston, Texas, and head of the human space exploration program of the 1960s, provides valuable insight into the origins and direction of Gemini during the mid-1960s.

Document 8
GEMINI PROGRAM

The Gemini program was designed to investigate in actual flight many of the critical situations which we would face later in the voyage of Apollo.

The spacecraft carried an onboard propulsion system for maneuvering in Earth orbit. A guidance and navigation system and a rendezvous radar were provided to permit astronauts to try out various techniques of rendezvous and docking with an Agena target vehicle. After docking, the astronauts could light off the Agena rocket for large changes in orbit, simulating the entry-into-lunar-orbit and the return-to-Earth burns of Apollo. Gemini was the first to use the controlled reentry system that was required for Apollo in returning from the Moon. It had hatches that could be opened and closed in space to permit extravehicular activity by astronauts, and fuel cells similar in purpose to those of Apollo to permit flights of long duration. The spacecraft was small by Apollo standards, carrying only two men in close quarters. However, the Titan II launch vehicle, which was the best available at that time, could not manage a larger payload.

A total of 10 manned flights were made in the Gemini program between March 1965 and November 1966. They gave us nearly 2000 man-hours in space and developed the rendezvous and docking techniques essential to Apollo. By burning the Agena rockets after docking, we were able to go to altitudes of more than 800 nautical miles and prove the feasibility of the precise space maneuvers essential to Apollo. Our first experience in EVA was obtained with Gemini and difficulties here early in the program paved the way for the smoothly working EVA systems used later on the Moon. The Borman and Lovell flight, Gemini VII, showed us that durations up to two weeks were possible without serious medical problems, and the later flights showed the importance of neutral buoyancy training in preparation of zero-gravity operations outside the spacecraft.

Gemini gave us the confidence we needed in complex space operations, and it was during this period that Chris Kraft and his team really made spaceflight operational. They devised superb techniques for flight management, and Mission Control developed to where it was really ready for the complex Apollo missions. Chris Kraft, Deke Slayton, head of the astronauts, and Dr. Berry, our head of Medical Operations, learned to work together as a team. Finally, the success of these operations and the high spaceflight activity kept public interest at a peak, giving our national leaders the broad supporting interest and general approval that made it possible to press ahead with a program of the scale of Apollo.

Source: Robert R. Gilruth, "I Believe We Should Go to the Moon," in Edgar M. Cortright, ed., *Apollo Expeditions to the Moon* (Washington, DC: NASA SP-350, 1975), p. 34.

DISASTER AND RECOVERY: THE *APOLLO 204* CAPSULE FIRE

On 27 January 1967, *Apollo-Saturn (AS) 204*, scheduled to be the first Apollo flight with astronauts aboard, was on the launch pad at Kennedy Space Center, Florida, moving through simulation tests with crewmembers Virgil I. "Gus" Grissom, Edward White, and Roger Chaffee. That evening, after several hours of work, a fire erupted in the pure oxygen atmosphere, and in a flash of flames engulfed the capsule. The astronauts died of asphyxiation. As the nation mourned, NASA went to work to determine the cause of the fire and to make alterations to the capsule to ensure the safety of future crews.

In the review process, it was found that questions about the capsule had been raised more than two years earlier. Late in 1965 Major General Samuel C. Phillips, Apollo program director at NASA headquarters, initiated a review of management of both the Apollo spacecraft and the *Saturn V* programs. This report gained added significance in the aftermath of the fire. NASA administrator James E. Webb was apparently unaware of its existence and was embarrassed by this on Capitol Hill during testimony about the accident.

Just as important as Phillips's report, investigations into the *Apollo 204* accident identified the direct reasons for the fire and made recommendations for capsule alterations based on these findings.

Document 9
NASA REVIEW TEAM REPORT (19 DECEMBER 1965)

I. Introduction

This is the report of the NASA's Management Review of North American Aviation Corporation management of Saturn II Stage (S-II) and Command and Service Module (CSM) programs. The Review was conducted as a result of the continual failure of NAA to achieve the progress required to support the objective of the Apollo Program.

The scope of the Review included an examination of the Corporate organization and its relationship to and influence on the activities of S&ID, the operating Division charged with the execution of the S-II and CSM programs. The Review also included examination of NAA off-site program activities at KSC [Kennedy Space Center] and MTF [Mississippi Test Facility].

The members of the Review team were specifically chosen for their experience with S&ID and their intimate knowledge of the S-II and CSM programs. The Review findings, therefore, are a culmination of the judgments of responsible government personnel directly involved with these programs. The team report represents an assessment of the contractor's performance and existing conditions affecting current and future progress, and recommends actions believed necessary to achieve an early return to the position supporting Apollo program objectives.

The Review was conducted from November 22 through December 6 [1965] and was organized into a Basic Team, responsible for over-all assessment of the contractor's activities and the relationships among his organizational elements and functions; and sub-teams who assessed the contractor's activities in the following areas:

Program Planning and Control (including Logistics)

Contracting, Pricing, Subcontracting, Purchasing

Engineering

Manufacturing

Reliability and Quality Assurance . . .

II. NAA's Performance to Date—Ability to Meet Commitments

At the start of the CSM and S-II Programs, key milestones were agreed upon, performance requirements established and cost plans developed. These were essentially commitments made by NAA to NASA. As the program progressed NASA has been forced to accept slippages in key milestone accomplishments, degradation in hardware performance, and increasing costs.

A. S-II

1. Schedules

. . . [K]ey performance milestones in testing, as well as end item hardware deliveries, have slipped continuously in spite of deletions of both hardware and test content. . . . [T]he delivery of the common bulkhead test article was rescheduled 5 times, for a total slippage of more than a year. . . . The total Apollo program was reoriented during this time, the S-II flight stages have remained behind schedules even after this reorientation.

2. Costs

The S-II cost picture . . . has been essentially a series of cost escalations with a bow wave of peak costs advancing steadily throughout the program life. Each annual projection has shown either the current or succeeding year to be the peak. NAA's estimate of the total 10 stage program has more than tripled. These increases have occurred despite the fact that there have been reductions in hardware.

3. Technical Performance

The S-II stage is still plagued with technical difficulties. . . . Welding difficulties, insulation bonding, continued redesign as a result of component failures during qualification are indicative of insufficiently aggressive pursuit of technical resolutions during the earlier phases of the program.

B. CSM [Command and Service Module]

1. Schedules

A history of slippages in meeting key CSM milestones [exists]. . . . The propulsion spacecraft, the systems integration spacecraft, and the spacecraft for the first development flight have each slipped more than six months. In addition, the first manned and the key environmental ground spacecraft have each slipped more than a year. These slippages have occurred in spite of the fact that schedule requirements have been revised a number of times, and seven articles, originally required for delivery by the end of 1965, have been eliminated. . . . The start of major testing in the ground test program has slipped from three to nine months in less than two years.

2. Costs

Analysis of spacecraft forecasted costs . . . reveals NAA has not been able to forecast costs with any reasonable degree of accuracy. The peak of the program cost has slipped 18 months in two years. In addition, NAA is forecasting that the total cost of the reduced spacecraft program will be greater than the cost of the previous planned program.

3. Technical Performance

Inadequate procedures and controls in bonding and welding, as well as inadequate master tooling, have delayed fabrication of airframes. In addition, there are still major development problems to be resolved. . . .

III. NASA Assessment—Probability of NAA Meeting Future Commitments

A. S-II

Today, after 4 ½ years and a little more than a year before first flight, there are still significant technical problems and unknowns affecting the stage. Manufacture is at least 5 months behind schedule. NAA's continued inability to meet internal objectives, as evidenced by 5 changes in the manufacturing plan in the last 3 months, clearly indicates that extraordinary effort will be required if the contractor is to hold the current position, let alone better it. . . . Failures in timely and complete engineering support, poor workmanship, and other conditions have also contributed to the current S-II situation. Factors which have caused these problems still exist. The two recent funding requirements exercises, with their widely different results, . . . leave little basis for confidence in the contractor's ability to accomplish the required work within the funds estimated. The team did not find significant indications of actions underway to build confidence that future progress will be better than past performance.

B. CSM

With the first unmanned flight spacecraft finally delivered to KSC, there are still significant problems remaining for Block I and Block II CSM's. Technical problems with electrical power capacity, service propulsion, structural integrity, weight growth, etc. have yet to be resolved. . . . Delayed and compromised ground and qualification test programs give us serious concern that fully qualified flight vehicles will not be available to support the lunar landing program. NAA's inability to meet spacecraft contract use deliveries has caused rescheduling of the total Apollo program. . . . While our management review indicated that some progress is being made to improve the CSM outlook, there is little confidence that NAA will meet its schedule and performance commitments within the funds available for this portion of the Apollo program.

IV. Summary Findings

Presented below is a summary of the team's views on those program conditions and fundamental management deficiencies that are impeding program progress and that require resolution by NAA to ensure that the CSM and S-II Programs regain the required program position. The detailed findings and recommendations of the individual sub-team reviews are in the Appendix to this report.

A. NAA performance on both programs is characterized by continued failure to meet committed schedule dates with required technical performance and within costs. There is no evidence of current improvement in NAA's management of these programs of the magnitude required to give confidence that NAA performance will improve at the rate required to meet established Apollo program objectives. . . .

Appendix:

Conclusions and Recommendations

. . . The findings are expressed frankly and result from the team's work in attempting to relate the end results we see in program conditions to fundamental causes for these conditions.

In most instances, recommendations for improvement accompany the findings. In some cases, problems are expressed for which the team has no specific recommendations, other than the need for attention and resolution by NAA.

It is not NASA's intent to dictate solutions to the deficiencies noted in this report. The solution to NAA's internal problems is both a prerogative and a responsibility of NAA Management, within the parameters of NASA's requirements as stated in the contracts. NASA does, however, fully expect objective, responsible, and timely action by NAA to correct the conditions described in this report.

It is recommended that the CSM incentive contract conversion proceed as now planned.

Incentivization of the S-II Program should be delayed until NASA is assured that the S-II Program is under control and a responsible proposal is received from the contractor.

Decision on a follow-on incentive contract for the CSM, beyond the present contract period, will be based on contractor performance.

It is recommended that NAA respond to NASA, by the end of January 1966, on the actions taken and planned to be taken to correct the conditions described in this report. At that time, NAA is also to certify the tasks, schedules, and resource requirements for the S-II and CSM Programs. . . .

Source: "NASA Review Team Report," attached to Major General Samuel C. Phillips, USAF, Apollo Program Director, to J. Leland Atwood, President, North American Aviation, Inc., 19 December 1965, NASA Historical Reference Collection, History Office, NASA Headquarters, Washington, DC.

Document 10
APOLLO 204 ACCIDENT BOARD REPORT (MARCH 1967)

Board Findings, Determinations and Recommendations

In this Review, the Board adhered to the principle that reliability of the Command Module and the entire system involved in its operation is a requirement common to both safety and mission success. Once the Command Module has left the earth's environment the occupants are totally dependent upon it for their safety. It follows that protection from fire as a hazard involves much more than quick egress. The latter has merit only during test periods on earth when the Command Module is being readied for its mission and not during the mission itself. The risk of fire must be faced; however, that risk is only one factor pertaining to the reliability of the Command Module that must receive adequate consideration. Design features and operating procedures that are intended to reduce the fire risk must not introduce other serious risks to mission success and safety.

1. Finding:

a. There was a momentary power failure at 23:30:55 GMT.

b. Evidence of several arcs was found in the post fire investigation.

c. No single ignition source of the fire was conclusively identified.

DETERMINATION:

The most probable initiator was an electrical arc in the sector between the −Y and +Z spacecraft axes. The exact location best fitting the total available information is near the floor in the lower forward section of the left-hand equipment bay where Environmental Control System (ECS) instrumentation power wiring leads into the area between the Environmental Control Unit panel. No evidence was discovered that suggested sabotage.

2. Finding:

a. The Command Module contained many types and classes of combustible material in areas contiguous to possible ignition sources.

b. The test was conducted with a 16.7 pounds per square inch absolute, 100 percent, oxygen atmosphere.

DETERMINATION:

The test conditions were extremely hazardous.

RECOMMENDATION:

The amount and location of combustible materials in the Command Module must be severely restricted and controlled.

3. Finding:

a. The rapid spread of fire caused an increase in pressure and temperature which resulted in rupture of the Command Module and creation of a toxic atmosphere. Death of the crew was from asphyxiation due to inhalation of toxic gases due to fire. A contributory cause of death was thermal burns.

b. Non-uniform distribution of carboxyhemoglobin was found by autopsy.

DETERMINATION:

Autopsy data leads to the medical opinion that unconsciousness occurred rapidly and that death followed soon thereafter.

4. Finding:

Due to internal pressure, the Command Module inner hatch could not be opened prior to rupture of the Command Module.

DETERMINATION:

The crew was never capable of effecting emergency egress because of the pressurization before rupture and their loss of consciousness soon after rupture.

RECOMMENDATION:

The time required for egress of the crew be reduced and the operations necessary for egress be simplified.

5. Finding:

Those organizations responsible for the planning, conduct and safety of this test failed to identify it as being hazardous. Contingency preparations to permit escape or rescue of the crew from an internal Command Module fire were not made.

a. No procedures for this type of emergency had been established either for the crew or for the spacecraft pad work team.

b. The emergency equipment located in the White Room and on the spacecraft work levels was not designed for the smoke condition resulting from a fire of this nature.

c. Emergency fire, rescue and medical teams were not in attendance.

d. Both the spacecraft work levels and the umbilical tower access arm contain features such as steps, sliding doors and sharp turns in the egress paths which hinder emergency operations.

DETERMINATION:

Adequate safety precautions were neither established nor observed for this test.

RECOMMENDATIONS:

a. Management continually monitor the safety of all test operations and assure the adequacy of emergency procedures.
b. All emergency equipment (breathing apparatus, protective clothing, deluge systems, access arm, etc.) be reviewed for adequacy.
c. Personnel training and practice for emergency procedures be given on a regular basis and reviewed prior to the conduct of a hazardous operation.
d. Service structures and umbilical towers be modified to facilitate emergency operations.

6. Finding:

Frequent interruptions and failures had been experienced in the overall communication system during the operations preceding the accident.

DETERMINATION:

The overall communication system was unsatisfactory.

RECOMMENDATIONS:

a. The Ground Communication System be improved to assure reliable communications between all test elements as soon as possible and before the next manned flight.
b. A detailed design review be conducted on the entire spacecraft communication system.

7. Finding:

a. Revisions to the Operational Checkout Procedure for the test were issued at 5:30 PM EST January 26, 1967 (209 pages) and 10:00 AM EST January 27, 1967 (4 pages).
b. Differences existed between the Ground Test Procedures and the In-Flight Check Lists.

DETERMINATION:

Neither the revision nor the differences contributed to the accident. The late issuance of the revision, however, prevented test personnel from becoming adequately familiar with the test procedure prior to its use.

RECOMMENDATIONS:

a. Test Procedures and Pilot's Checklists that represent the actual Command Module configuration be published in final form and reviewed early enough to permit adequate preparation and participation of all test organization.

b. Timely distribution of test procedures and major changes be made a constraint to the beginning of any test.

8. Finding:

The fire in Command Module 012 was subsequently simulated closely by a test fire in a full-scale mock-up.

DETERMINATION:

Full-scale mock-up fire tests can be used to give a realistic appraisal of fire risks in flight-configured spacecraft.

RECOMMENDATION:

Full-scale mock-ups in flight configuration be tested to determine the risk of fire.

9. Finding:

The Command Module Environmental Control System design provides a pure oxygen atmosphere.

DETERMINATION:

This atmosphere presents severe fire hazards if the amount and location of combustibles in the Command Module are not restricted and controlled.

RECOMMENDATIONS:

a. The fire safety of the reconfigured Command Module be established by full-scale mock-up tests.

b. Studies of the use of a diluent gas be continued with particular reference to assessing the problems of gas detection and control and the risk of additional operations that would be required in the use of a two gas atmosphere.

10. Finding:

Deficiencies existed in Command Module design, workmanship and quality control, such as:

a. Components of the Environmental Control System installed in Command Module 012 had a history of many removals and of technical difficulties including regulator failures, line failures and Environmental Control Unit failures. The design and installation features of the Environmental Control Unit makes removal or repair difficult.

b. Coolant leakage at solder joints has been a chronic problem.

c. The coolant is both corrosive and combustible.

d. Deficiencies in design, manufacture, installation, rework and quality control existed in the electrical wiring.

e. No vibration test was made of a complete flight-configured spacecraft.

f. Spacecraft design and operating procedures currently require the disconnecting of electrical connections while powered.

g. No design features for fire protection were incorporated.

DETERMINATION:

These deficiencies created an unnecessarily hazardous condition and their continuation would imperil any future Apollo operations.

RECOMMENDATIONS:

a. An in-depth review of all elements, components and assemblies of the Environmental Control System be conducted to assure its functional and structural integrity and to minimize its contribution to fire risk.

b. Present design of soldered joints in plumbing be modified to increase integrity or the joints be replaced with a more structurally reliable configuration.

c. Deleterious effects of coolant leakage and spillage be eliminated.

d. Review of specifications be conducted, 3-dimensional Jigs be used in manufacture of wire bundles and rigid inspection at all stages of wiring design, manufacture and installation be enforced.

e. Vibration tests be conducted of a flight-configured spacecraft.

f. The necessity for electrical connections or disconnections with power on within the crew compartment be eliminated.

g. Investigation be made of the most effective means of controlling and extinguishing a spacecraft fire. Auxiliary breathing oxygen and crew protection from smoke and toxic fumes be provided. . . .

Source: Apollo 204 Accident Board Report Board, "Findings, Determinations and Recommendations," March 1967, copy in NASA Historical Reference Collections, History Office, NASA Headquarters, Washington, DC.

TASK ACCOMPLISHED: THE CULMINATION OF APOLLO

In the aftermath of the tragic *Apollo 204* capsule fire in 1967, the U.S. goal of reaching the moon before the end of the decade seemed in jeopardy. It was almost twenty months after the fire when astronauts finally were launched into orbit (in October 1968) aboard an Apollo spacecraft. The success of this test flight led to a rapid succession of ever more sophisticated missions in the winter of 1968–1969. Notably, the dramatic circumlunar mission of *Apollo 8* on 21–27 December 1968 heralded the first lunar landing mission the next summer.

The two astronauts of *Apollo 11* who eventually set foot on the lunar surface on 20 July 1969, Neil A. Armstrong and Edwin E. "Buzz" Aldrin, called it "magnificent desolation." The first document in this section is a transcript of the radio transmissions of the landing and Armstrong's first venture onto the lunar surface. The "CC" in the transcript is Houston Mission Control; "CDR" is Neil Armstrong and "LMP" is Buzz Aldrin.

The second document in this section is the president's congratulation of the NASA team on their successful completion of Apollo.

Document 11
APOLLO 11 MOON/EARTH TRANSMISSIONS
(20 JULY 1969)

Time	Position	Comment
04:06:45:52	CC	We copy you down, Eagle [the name of the lunar module].
04:06:45:57	CDR	Houston, Tranquility Base here.
04:06:45:59	CDR	THE EAGLE HAS LANDED.
04:06:46:04	CC	Roger, Tranquility. We copy you on the ground. You got a bunch of guys about to turn blue. We're breathing again. Thanks a lot.
04:06:46:16	CDR	Thank you.
04:06:46:18	CC	You're looking good here.
04:06:46:23	CDR	Okay. We're going to be busy for a minute. . . .
04:06:46:38	LMP	Very smooth touchdown. . . .
04:06:47:03	LMP	Okay. It looks like we're venting the oxidizer now. . . .
04:06:47:09	CC	Eagle you are STAY for T1 [one day on moon].
04:06:47:12	CDR	Roger. Understand, STAY for T1. . . .
04:13:23:38	CDR	[After suiting up and exiting the lunar module (LM), Armstrong was ready to descend to the moon's surface.] I'm at the foot of the ladder. The LM footpads are only depressed in the surface about 1 or 2 inches, although the surface appears to be very, very fine grained, as you get close to it. It's almost like a power. Down there, it's very fine.
04:13:23:43	CDR	I'm going to step off the LM now.
04:13:24:48	CDR	THAT'S ONE SMALL STEP FOR MAN, ONE GIANT LEAP FOR MANKIND.

04:13:24:58	CDR	And the—the surface is fine and powdery. I can—I can pick it up loosely with my toe. It does adhere in fine layers like powdered charcoal to the sole and sides of my boots. I only go in a small fraction of an inch, maybe an eighth of an inch, but I can see the footprints of my boots and the treads in the fine, sandy particles.
04:13:25:30	CC	Neil, this is Houston. We're copying.
04:13:25:45	CDR	There seems to be no difficulty in moving around as we suspected. It's even perhaps easier than the simulations at one-sixth g that we performed in the various simulations on the ground. It's actually no trouble to walk around. Okay. The descent engine did not leave a crater of any size. It has about 1 foot clearance on the ground. We're essentially on a very level place here. I can see some evidence of rays emanating from the descent engine, but a very insignificant amount. . . .

Source: NASA, Manned Spacecraft Center, "Apollo 11 Technical Air-to-Ground Voice Transcription," July 1969, pp. 317, 377, Apollo 11 Files, NASA Historical Reference Collection, History Office, NASA Headquarters, Washington, DC.

Document 12
THE PRESIDENT'S CONGRATULATION
(21 MARCH 1972)

March 24, 1972

Fellow Members of the Apollo Team:

I have received the following letter from President Nixon in which he said he wanted the Apollo Team to know how much this nation values the work we have done and are doing in the Apollo Program. The letter was addressed to me but the President's words were really addressed to each of you.

I am pleased to pass along the President's words which each of you has done so much to earn.

Sincerely,
[signed]
Rocco A. Petrone
Apollo Program Director

Dear Dr. Petrone:

As we approach the final countdown for Apollo 16, I want you and all the men and women of Apollo to know how much this nation values your splendid efforts. The moon flight program has captured the imagination of our times as has no other human endeavor. You and your team have, in fact, written the first chapter in the history of man's exploration of space, and all future achievements must credit all of you for having blazed the path.

Countless people throughout the world will soon be sharing with you the excitement of Apollo 16's voyage, and I know I speak for all of them in conveying to you my warmest best wishes for a safe and successful flight. Good luck!

Sincerely,
[signed]
Richard Nixon

Source: Rocco A. Petrone, NASA Apollo Program Manager, to Apollo Program Team, 24 March 1972, Apollo Collection, NASA Historical Reference Collection, History Office, NASA Headquarters, Washington, DC.

RECOLLECTIONS FROM THE CREW OF *APOLLO 11*

The following mission account makes use of crew members' own words, from transcript of astronaut Michael Collins, who remained in the *Apollo 11* command module while Neil A. Armstrong and Edwin E. "Buzz" Aldrin Jr. visited the surface of the moon on 20 July 1969.

Document 13
"THE EAGLE HAS LANDED" (20 JULY 1969)

COLLINS: I am lastingly thankful that I have flown before, and that this period of waiting atop a rocket is nothing new. I am just as tense this time, but the tenseness comes mostly from an appreciation of the enormity of our undertaking rather than from the unfamiliarity of the situation. I am far from certain that we will be able to fly the mission as planned. I think we will escape with our skins, or at least I will escape with mine, but I wouldn't give better than even odds on a successful landing and return. There are just too many things that can go wrong. Fred Haise [the backup astronaut who had checked command-module switch positions] has run through a checklist 417 steps long, and I have merely a half dozen minor chores to take care of—nickel and dime stuff. In between switch throws I have plenty of time to think, if not daydream. Here I am, a white male, age thirty-eight, height 5 feet 11 inches, weight 165 pounds, salary

$17,000 per annum, resident of a Texas suburb, with black spot on my roses, state of mind unsettled, about to be shot off to the Moon. Yes, to the Moon.

At the moment, the most important control is over on Neil's side, just outboard of his left knee. It is the abort handle, and now it has power to it, so if Neil rotates it 30 [degrees] counterclockwise, three solid rockets above us will fire and yank the CM free of the service module and everything below it. It is only to be used in extremes. A large bulky pocket has been added to Neil's left suit leg, and it looks as though if he moves his leg slightly, it's going to snag on the abort handle. I quickly point this out to Neil, and he grabs the pocket and pulls it as far over to the inside of his thigh as he can, but it still doesn't look secure to either one of us. Jesus, I can see the headlines now: " 'MOONSHOT FALLS INTO OCEAN.' Mistake by crew, program officials intimate. Last transmission from Armstrong prior to leaving the pad reportedly was 'Oops.' "

ARMSTRONG: The flight started promptly, and I think that was characteristic of all events of the flight. The Saturn gave us one magnificent ride, both in Earth orbit and on a trajectory to the Moon. Our memory of that differs little from the reports you have heard from the previous Saturn V flights.

ALDRIN: For the thousands of people watching along the beaches of Florida and the millions who watched on television, our liftoff was ear shattering. For us there was a slight increase in the amount of background noise, not at all unlike the sort one notices taking off in a commercial airliner, and in less than a minute we were traveling ahead of the speed of sound.

COLLINS: This beast is best felt. Shake, rattle, and roll! We are thrown left and right against our straps in spasmodic little jerks. It is steering like crazy, like a nervous lady driving a wide car down a narrow alley, and I just hope it knows where it's going, because for the first ten seconds we are perilously close to that umbilical tower.

ALDRIN: A busy eleven minutes later we were in Earth orbit. The Earth didn't look much different from the way it had during my first flight, and yet I kept looking at it. From space it has an almost benign quality. Intellectually one could realize there were wars underway, but emotionally it was impossible to understand such things. The thought reoccurred that wars are generally fought for territory or are disputes over borders; from space the arbitrary borders established on Earth cannot be seen. After one and a half orbits a preprogrammed sequence fired the Saturn to send us out of Earth orbit and on our way to the Moon. . . .

Fourteen hours after liftoff, at 10:30 P.M. Houston time, the three astronauts fastened covers over the windows of the slowly rotating command module and went to sleep. Days 2 and 3 were devoted to housekeeping chores, a small midcourse velocity correction, and televised transmissions back to Earth. . . .

The Most Awesome Sphere

COLLINS: Day 4 has a decidedly different feel to it. Instead of nine hours' sleep, I get seven—and fitful ones at that. Despite our concentrated effort to conserve our energy on the way to the Moon, the pressure is overtaking us (or me at least), and I feel that all of us are aware that the honeymoon is over and we are about to lay our little pink bodies on the line. Our first shock comes as we stop our spinning motion and swing ourselves around so as to bring the Moon into view. We have not been able to see the Moon for nearly a day now, and the change is electrifying. The Moon I have known all my life, that two-dimensional small yellow disk in the sky, has gone away somewhere, to be replaced by the most awesome sphere I have ever seen. To begin with it is huge, completely filling our window. Second, it is three-dimensional. The belly of it bulges out toward us in such a pronounced fashion that I almost feel I can reach out and touch it. To add to the dramatic effect, we can see the stars again. We are in the shadow of the Moon now, and the elusive stars have reappeared.

As we ease around on the left side of the Moon, I marvel again at the precision of our path. We have missed hitting the Moon by a paltry 300 nautical miles, at a distance of nearly a quarter of a million miles from Earth, and don't forget that the Moon is a moving target and that we are racing through the sky just ahead of its leading edge. When we launched the other day the Moon was nowhere near where it is now; it was some 40 degrees of arc, or nearly 200,000 miles, behind where it is now, and yet those big computers in the basement in Houston didn't even whimper but belched out super-accurate predictions.

As we pass behind the Moon, we have just over eight minutes to go before the burn. We are super-careful now, checking and rechecking each step several times. When the moment finally arrives, the big engine instantly springs into action and reassuringly plasters us back in our seats. The acceleration is only a fraction of one G but it feels good nonetheless. For six minutes we sit there peering intent as hawks at our instrument panel, scanning the important dials and gauges, making sure that the proper thing is being done to us. . . .

ALDRIN: The second burn to place us in closer circular orbit of the Moon, the orbit from which Neil and I would separate from the Columbia and continue on to the Moon, was critically important. . . . [T]wo hours after our first lunar orbit we made our second burn, in an atmosphere of nervous and intense concentration. It, too, worked perfectly.

Asleep in Lunar Orbit

We began preparing the LM. It was scheduled to take three hours, but because I had already started the checkout, we were completed a half hour ahead of schedule. Reluctantly we returned to the Columbia as planned. Our fourth night we were to sleep in lunar orbit. Although it was not in the flight plan, before covering the windows and dousing the lights, Neil and I carefully prepared all the equipment and clothing we would need in the morning, and mentally ran through the many procedures we would follow.

COLLINS: "Apollo 11, Apollo 11, good morning from the Black Team." Could they be talking to me? It takes me twenty seconds to fumble for the microphone button and answer groggily, I guess I have only been asleep five hours or so; I had a tough time getting to sleep, and now I'm having trouble waking up. Neil, Buzz, and I all putter about fixing breakfast and getting various items ready for transfer into the LM. [Later.] I stuff Neil and Buzz into the LM along with an armload of equipment. Now I have to do the tunnel bit again, closing hatches, installing drogue [parachute] and probe, and disconnecting the electrical umbilical. I am on the radio constantly now, running through an elaborate series of joint checks with Eagle. I check progress with Buzz: "I have five minutes and fifteen seconds since we started. Attitude is holding very well." "Roger, Mike, just hold it a little bit longer." "No sweat, I can hold it all day. Take your sweet time. How's the czar over there? He's so quiet." Neil chimes in, "Just hanging on—and punching." Punching those computer buttons, I guess he means. "All I can say is, beware the revolution," and then, getting no answer, I formally bid them goodbye. "You cats take it easy on the lunar surface. . . ." "O.K., Mike," Buzz answers cheerily, and I throw the switch which releases them. With my nose against the window and the movie camera churning away, I watch them go. When they are safely clear of me, I inform Neil, and he begins a slow pirouette in place, allowing me a look at his outlandish machine and its four extended legs. "The Eagle has wings," Neil exults.

It doesn't look like any eagle I have ever seen. It is the weirdest-looking contraption ever to invade the sky, floating there with its legs awkwardly jutting out above a body which has neither symmetry nor grace. I make sure all four landing gears are down and locked, report that fact, and then lie a little, "I think you've got a fine-looking flying machine there, Eagle, despite the fact you're upside down." "Somebody's upside down," Neil retorts. "O.K., Eagle. One minute . . . you guys take care." Neil answers, "See you later." I hope so. When the one minute is up, I fire my thrusters precisely as planned and we begin to separate, checking distances and velocities as we go. This burn is a very small one, just to give Eagle some breathing room. From now on it's up to them, and they will make two separate burns in reaching the lunar surface. The first one will serve to drop Eagle's perilune [closest approach in orbit] to fifty thousand feet. Then, when they reach this spot over the eastern edge of the Sea of Tranquility, Eagle's descent engine will be fired up for the second and last time, and Eagle will lazily arc over into a 12-minute computer-controlled descent to some point at which Neil will take over for a manual landing.

ALDRIN: We were still 60 miles above the surface when we began our first burn. Neil and I were harnessed into the LM in a standing position. [Later] at precisely the right moment the engine ignited to begin the 12-minute powered descent. Strapped in by the system of belts and cables not unlike shock absorbers, neither of us felt the initial motion. We looked quickly at the computer to make sure we were actually functioning as planned. After 26 seconds the engine went to full throttle and the motion became noticeable. Neil watched his instruments while I looked at our primary computer and compared it with our second computer, which was part of our abort guidance system. . . .

A Yellow Caution Light

At six thousand feet above the lunar surface a yellow caution light came on and we encountered one of the few potentially serious problems in the entire flight, a problem which might have caused us to abort, had it not been for a man on the ground who really knew his job.

COLLINS: At five minutes into the burn, when I am nearly directly overhead, Eagle voices its first concern. "Program Alarm," barks Neil, "It's a 1202." What the hell is that? I don't have the alarm numbers memorized for my own computer, much less for the LM's. I jerk out my own checklist and start thumbing through it, but before I can find 1202, Houston says, "Roger, we're GO on that alarm." No problem, in other words. My checklist says 1202 is an "executive overflow," meaning simply that the computer has been called upon to do too many things at once and is forced to postpone some of them. A little farther along, at just three thousand feet above the surface, the computer flashes 1201, another overflow condition, and again the ground is superquick to respond with reassurances.

ALDRIN: Back in Houston, not to mention on board the Eagle, hearts shot up into throats while we waited to learn what would happen. We had received two of the caution lights when Steve Bales, the flight controller responsible for LM computer activity, told us to proceed, through Charlie Duke, the capsule communicator. We received three or four more warnings but kept on going. When Mike, Neil, and I were presented with Medals of Freedom by President Nixon, Steve also received one. He certainly deserved it, because without him we might not have landed. . . .

ARMSTRONG: Once [we] settled on the surface, the dust settled immediately and we had an excellent view of the area surrounding the LM.

We saw a crater surface, pockmarked with craters up to 15, 20, 30 feet, and many smaller craters down to a diameter of 1 foot and, of course, the surface was very fine-grained. There were a surprising number of rocks of all sizes.

A number of experts had, prior to the flight, predicted that a good bit of difficulty might be encountered by people due to the variety of strange atmospheric and gravitational characteristics. This didn't prove to be the case and after landing we felt very comfortable in the lunar gravity. It was, in fact, in our view preferable both to weightlessness and to the Earth's gravity.

When we actually descended the ladder it was found to be very much like the lunar-gravity simulations we had performed here on Earth. No difficulty was encountered in descending the ladder. The last step was about 3 ½ feet from the surface, and we were somewhat concerned that we might have difficulty in reentering the LM at the end of our activity period. So we practiced that before bringing the camera down.

ALDRIN: We opened the hatch and Neil, with me as his navigator, began backing out of the tiny opening. It seemed like a small eternity before I heard Neil say, "That's one small step for man . . . one giant leap for mankind." In less than fifteen minutes I was backing awkwardly out of the hatch and onto the surface to join Neil, who, in the tradition of all tourists, had his camera ready to photograph my arrival.

I felt buoyant and full of goose pimples when I stepped down on the surface. I immediately looked down at my feet and became intrigued with the peculiar properties of the lunar dust. If one kicks sand on a beach, it scatters in numerous directions with some grains traveling farther than others. On the Moon the dust travels exactly and precisely as it goes in various directions, and every grain of it lands nearly the same distance away.

ARMSTRONG: There were a lot of things to do, and we had a hard time getting them finished. . . . The primary difficulty was just far too little time to do the variety of things we would have liked. We had the problem of the five-year-old boy in a candy store.

ALDRIN: I took off jogging to test my maneuverability. The exercise gave me an odd sensation and looked even more odd when I later saw the films of it. With bulky suits on, we seemed to be moving in slow motion. I noticed immediately that my inertia seemed much greater. Earth-bound, I would have stopped my run in just one step, but I had to use three or four steps to sort of wind down. My Earth weight, with the big backpack and heavy suit, was 360 pounds. On the Moon I weighed only 60 pounds [Even though I had on my lunar spacesuit].

At one point I remarked that the surface was "Beautiful, beautiful. Magnificent desolation." I was struck by the contrast between the starkness of the shadows and the desert-like barrenness of the rest of the surface. It ranged from dusty gray to light tan and was unchanging except for one startling sight: our LM sitting there with its black, silver, and bright yellow-orange thermal coating shining brightly in the otherwise colorless landscape. I had seen Neil in his suit thousands of times before, but on the Moon the unnatural whiteness of it seemed unusually brilliant. We could also look around and see the Earth, which, though much larger than the Moon the Earth was seeing, seemed small—a beckoning oasis shining far away in the sky.

As the sequence of lunar operations evolved, Neil had the camera most of the time, and the majority of pictures taken on the Moon that include an astronaut are of me. It wasn't until we were back on Earth and in the Lunar Receiving Laboratory looking over the pictures that we realized there were few pictures of Neil. My fault, perhaps, but we had never simulated this in our training.

Coaxing the Flag to Stand

During a pause in experiments, Neil suggested we proceed with the flag. It took both of us to set it up and it was nearly a disaster. Public Relations obviously needs practice just as everything else does. A small telescoping arm was attached to the flagpole to keep the flag extended and perpendicular. As hard as we tried, the telescope wouldn't fully extend. Thus the flag, which should have been flat, had its own unique permanent wave. Then to our dismay the staff of the pole wouldn't go far enough into the lunar surface to support itself in an upright position. After much struggling we finally coaxed it to remain upright, but in a most precarious position. I dreaded the possibility of the American flag collapsing into the lunar dust in front of the television camera.

COLLINS: [On his fourth orbital pass above.] "How's it going?" "The EVA is progressing beautifully. I believe they're setting up the flag now." Just let things keep going that way, and no surprises, please. Neil and Buzz sound good, with no huffing and puffing to indicate they are overexerting themselves. But one surprise at least is in store. Houston comes on the air, not the slightest bit ruffled, and announces that the President of the United States would like to talk to Neil and Buzz. "That would be an honor," says Neil, with characteristic dignity.

The President's voice smoothly fills the air waves with the unaccustomed cadence of the speechmaker, trained to convey inspiration, or at least emotion, instead of our usual diet of numbers and reminders. "Neil and Buzz, I am talking to you by telephone from the Oval Office at the White House, and this certainly has to be the most historic telephone call ever made. . . . Because of what you have done, the heavens have become a part of man's world. As you talk to us from the Sea of Tranquility, it inspires us to redouble our efforts to bring peace and tranquility to Earth. . . ." My God, I never thought of all this bringing peace and tranquility to anyone. As far as I am concerned, this voyage is fraught with hazards for the three of us—and especially two of us—and that is about as far as I have gotten in my thinking.

Neil, however, pauses long enough to give as well as he receives. "It's a great honor and privilege for us to be here, representing not only the United States but men of peace of all nations, and with interest and a curiosity and a vision for the future." [Later.] Houston cuts off the White House and returns to business as usual, with a long string of numbers for me to copy for future use. My God, the juxtaposition of the incongruous—roll, pitch, and yaw; prayers, peace, and tranquility. What will it be like if we really carry this off and return to Earth in one piece, with our boxes full of rocks and our heads full of new perspectives for the planet? I have a little time to ponder this as I zing off out of sight of the White House and the Earth.

ALDRIN: . . . Before beginning liftoff procedures [we] settled down for our fitful rest. We didn't sleep much at all. Among other things we were elated—and also cold. Liftoff from the Moon, after a stay totaling twenty-one hours, was exactly on schedule and fairly uneventful. The ascent stage of the LM separated, sending out a shower of brilliant insulation particles which had been ripped off from the thrust of the ascent engine. There was no time to sightsee. I was concentrating on the computers, and Neil was studying the attitude indicator, but I looked up long enough to see the flag fall over. . . . Three hours and ten minutes later we were connected once again with the Columbia.

COLLINS: . . . [After docking.] It's time to hustle down into the tunnel and remove hatch, probe, and drogue, so Neil and Buzz can get through. Thank God, all the claptrap works beautifully in this its final workout. The probe and drogue will stay with the LM and be abandoned with it, for we will have no further need of them and don't want them cluttering up the command module. The first one through is Buzz, with a big smile on his face. I grab his head, a hand on each temple, and am about to give him a smooch on the forehead, as a parent might greet an errant child; but then, embarrassed, I think better of it and grab his hand, and then Neil's. We cavort about a little bit, all smiles and giggles over our success, and then it's back to work as usual.

Source: Michael Collins and Edwin E. Aldrin Jr., "The Eagle Has Landed," in Edgar M. Cortright, ed., *Apollo Expeditions to the Moon* (Washington, DC: NASA SP-350, 1975), pp. 203–222, excerpts.

THE DECISION TO BUILD THE SPACE SHUTTLE

Within a short time of taking office in January 1969, President Richard M. Nixon appointed a Space Task Group to plan for post-Apollo space exploration. Although the Space Task Group recommended a grandiose space station, a lunar colony, and a mission to Mars, Nixon refused to endorse any of these efforts. Even if the most ambitious parts of the long-range space plan could not be funded in the tight fiscal environment of the early 1970s, Nixon was convinced to approve a reusable space shuttle. But not without some hesitancy.

A breakthrough on the space shuttle decision came on 12 August 1971 when Caspar W. Weinberger, deputy director of the Office of Management and Budget (OMB), wrote a memorandum to the president arguing that "there is real merit to the future of NASA, and to its proposed programs" and suggesting that Nixon approve the start-up of space shuttle development. In a handwritten scrawl on Weinberger's memo, Richard Nixon indicated "I agree with Cap."

These documents set the stage for a formal announcement by President Nixon on 5 January 1972 to build the space shuttle, as detailed in the following memorandum by NASA deputy administrator George M. Low and the statement by the president. The NASA leaders were told in a 3 January 1972 meeting that the White House had decided to approve the development of a partially reusable shuttle. The next day, Low and Administrator James C. Fletcher flew to California to meet on 5 January with President Nixon, who was at the Western White House in San Clemente, for a discussion of the shuttle project. After the meeting with the president, the White House announced approval of the shuttle to the press, and Fletcher and Low answered questions about the project.

Document 14
MEMORANDUM FOR THE PRESIDENT (12 AUGUST 1971)

From: Caspar W. Weinberger
Via: George P. Shultz
Subject: Future of NASA

Present tentative plans call for major reductions or changes in NASA, by eliminating the last two Apollo flights (16 and 17), and eliminating or sharply reducing the balance of the Manned Space Program (Skylab and Space Shuttle) and many remaining NASA programs.

I believe this would be a mistake.

(1) The real reason for sharp reductions in the NASA budget is that NASA is entirely in the 28% of the budget that is controllable. In short we cut it because it is cuttable, not because it is doing a bad job or an unnecessary one.

(2) We are being driven, by the uncontrollable items, to spend more and more on programs that offer no real hope for the future: Model Cities, OEO [Orbiting Earth Observatory], Welfare, interest on National Debt, unemployment compensation, Medicare, etc. Of course, some of these have to be continued, in one form or another, but essentially they are

programs, not of our choice, designed to repair mistakes of the past, not of our making.

(3) We do need to reduce the budget, in my opinion, but we should not make all our reduction decisions on the basis of what is reducible, rather than on the merits of individual programs.

(4) There is real merit to the future of NASA, and to its proposed programs. The Space Shuttle and NERVA particularly offer the opportunity, among other things, to secure substantial scientific fallout for the civilian economy at the same time that large numbers of valuable (and hard-to-employ-elsewhere) scientists and technicians are kept at work on projects that increase our knowledge of space, our ability to develop for lower cost space exploration, travel, and to secure, through NERVA, twice the existing propulsion efficiency of our rockets.

It is very difficult to re-assemble the NASA teams should it be decided later, after major stoppages, to re-start some of the long-range programs.

(5) Recent Apollo flights have been very successful from all points of view. Most important is the fact that they give the American people a much needed lift in spirit (and the people of the world an equally needed look at American superiority). Announcement now, or very shortly, that we were cancelling Apollo 16 and 17 (an announcement we would have to make very soon if any real savings are to be realized) would have a very bad effect, coming so soon after Apollo 15's triumph. It would be confirming in some respects, a belief that I fear is gaining credence at home and abroad: That our best years are behind us, that we are turning inward, reducing our defense commitments, and voluntarily starting to give up our super-power status, and our desire to maintain world superiority.

America should be able to afford something besides increased welfare, programs to repair our cities, or Appalachian relief and the like.

(6) I do not propose that we necessarily fund all NASA seeks—only that . . . we *are* going to fund space shuttles, NERVA, or other major, future NASA activities. . . .

[signed]
Caspar W. Weinberger

Source: Caspar W. Weinberger, Memorandum for the President, via George Shultz, "Future of NASA," 12 August 1971, White House, Richard M. Nixon, President, 1968–1971 File, NASA Historical Reference Collection, History Office, NASA Headquarters, Washington, DC.

Document 15
MEETING WITH THE PRESIDENT ON 5 JANUARY 1972

Memorandum for the Record

12 January 1972

Jim Fletcher and I met with the President and John Ehrlichman for approximately 40 minutes to discuss the space shuttle. During the course of the discussion, the President either made or agreed with the following points:

1. *The Space Shuttle.* The President stated that we should stress civilian applications but not to the exclusion of military applications. We should not hesitate to mention the military applications as well. He was interested in the possibility of routine operations and quick reaction times, particularly as these would apply to problems of natural disasters, such as earthquakes or floods. When Dr. Fletcher mentioned a future possibility of collecting solar power in orbit and beaming it down to earth, the President indicated that these kinds of things tend to happen much more quickly than we now expect and that we should not hesitate to talk about them now. He was also interested in the nuclear waste disposal possibilities. The President liked the fact that ordinary people would be able to fly in the shuttle, and that the only requirement for a flight would be that there is a mission to be performed. He also reiterated his concern for preserving the skills of the people in the aerospace industry.

In summary, the President said that even though we now know of many things that the shuttle will be able to do, we should realize that it will open up entirely new fields when we actually have the capability that the shuttle will provide. The President wanted to know if we thought the shuttle was a good investment and, upon receiving our affirmative reply, requested that we stress the fact that the shuttle is not a "$7 billion toy," that it is indeed useful, and that it is a good investment in that it will cut operations costs by a factor of 10. But he indicated that even if it were not a good investment, we would have to do it anyway, because space flight is here to stay. Men are flying in space now and will continue to fly in space, and we'd best be part of it.

2. *International Cooperation.* The President said that he is most interested in making the space program a truly international program and that he had previously expressed that interest. He wanted us to stress international cooperation and participation for *all* nations. He said that he was disappointed that we had been unable to fly foreign astronauts on Apollo, but understood the reasons for our inability to do so. He understood that foreign astronauts of all nations could fly in the shuttle and appeared to be particularly interested in Eastern European participation in the flight program. However, in connection with international cooperation, he is not only interested in flying foreign astronauts, but also in other types of meaningful participation, both in experiments and even in space hardware development. . . .

The president asked John Ehrlichman to mention both the international aspects of the shuttle and the USSR docking possibilities to Henry Kissinger.

[signed]
George M. Low

Source: George M. Low, Deputy Administrator, NASA, Memorandum for the Record, "Meeting with the President on January 5, 1972," 12 January 1972, NASA Historical Reference Collection, History Office, NASA Headquarters, Washington, DC.

Document 16
STATEMENT BY THE PRESIDENT (5 JANUARY 1972)

I have decided today that the United States should proceed at once with the development of an entirely new type of space transportation system designed to help transform the space frontier of the 1970s into familiar territory, easily accessible for human endeavor in the 1980s and '90s.

This system will center on a space vehicle that can shuttle repeatedly from earth to orbit and back. It will revolutionize transportation into near space, by routinizing it. It will take the astronomical costs out of astronautics. In short, it will go a long way toward delivering the rich benefits of practical space utilization and the valuable spinoffs from space efforts into the daily lives of Americans and all people.

The new year 1972 is a year of conclusion for America's current series of manned flights to the moon. Much is expected from the two remaining Apollo missions—in fact, their scientific results should exceed the return from all the earlier flights together. Thus they will place a fitting capstone on this vastly successful undertaking. But they also bring us to an important decision point—a point of assessing what our space horizons are as Apollo ends, and of determining where we go from here.

In the scientific arena, the past decade of experience has taught us that spacecraft are an irreplaceable tool for learning about our near-earth space environment, the moon, and the planets, besides being an important aid to our studies of the sun and stars. In utilizing space to meet needs on earth, we have seen the tremendous potential of satellites for intercontinental communications and world-wide weather forecasting. We are gaining the capability to use satellites as tools in global monitoring and management of natural resources, in agricultural applications, and in pollution control. We can foresee their use in guiding airliners across the oceans and in bringing televised education to wide areas of the world.

However, all these possibilities, and countless others with direct and dramatic bearing on human betterment, can never be more than fractionally realized so long as every single trip from earth to orbit remains a matter of special effort and staggering expense. This is why commitment to the space shuttle program is the right next step for America to take, in moving out from our present beachhead in the sky to achieve a real working presence in space—because the space shuttle will give us routine access to space by sharply reducing costs in dollars and preparation time.

The new system will differ radically from all existing booster systems, in that most of this new system will be recovered and used again and again—up to 100 times. The resulting economies may bring operating costs down as low as one-tenth of those for present launch vehicles.

The resulting changes in modes of flight and re-entry will make the ride safer and less demanding for the passengers, so that men and women with work to do in space can "commute" aloft, without having to spend years in training for the skills and rigors of old style space flight. As scientists and technicians are actually able to accompany their instruments into space, limiting boundaries between our manned and unmanned space programs will disappear. Development of new space applications will be able to proceed much faster. Repair or servicing of satellites in space will become possible, as will delivery of valuable payloads from orbit back to earth.

The general reliability and versatility which the shuttle offers seems likely to establish it quickly as the workhorse of our whole space effort, taking the place of all present launch vehicles except the very smallest and very largest.

NASA and many aerospace companies have carried out extensive design studies for the shuttle. Congress has reviewed and approved this effort. Preparation is now sufficient for us to commence the actual work of construction with full confidence of success. In order to minimize technical and economic risks, the space agency will continue to take a cautious evolutionary approach in the development of this new system. Even so, by moving ahead at this time, we can have the shuttle in manned flight by 1978, and operational a short time later.

It is also significant that this major new national enterprise will engage the best efforts of thousands of highly skilled workers and hundreds of contractor firms over the next several years. The amazing "technology explosion" that has swept this country in the years since we ventured into space should remind us that robust activity in the aerospace industry is healthy for everyone—not just in jobs and income, but in the extension of our capabilities in every direction. The continued preeminence of America and American industry in the aerospace field will be an important part of the shuttle's "payload."

Views of the earth from space have shown us how small and fragile our home planet truly is. We are learning the imperatives of universal brotherhood and global ecology—learning to think and act as guardians of one tiny blue and green island in the trackless oceans of the universe. This new program will give more people more access to the liberating perspectives of space, even as it extends our ability to cope with physical challenges of earth and broadens our opportunities for international cooperation in low-cost, multi-purpose space missions.

"We must sail sometimes with the wind and sometimes against it," said

Oliver Wendell Holmes, "but we must sail, and not drift, nor lie at anchor." So with man's epic voyage into space—a voyage the United States of America has led and still shall lead.

Source: White House Press Secretary, "The White House, Statement by the President," 5 January 1972, Richard M. Nixon Presidential Files, NASA Historical Reference Collection, History Office, NASA Headquarters, Washington, DC.

THE *CHALLENGER* ACCIDENT

The 28 January 1986 explosion of the space shuttle *Challenger* was one of the most traumatic events in recent American history. This section provides several documents exploring this important accident. The first is a transcript of the *Challenger*'s operational recorder voice tape. It reveals the comments of Commander Francis R. Scobee, Pilot Michael J. Smith, Mission Specialist 1 Ellison S. Onizuka, and Mission Specialist 2 Judith A. Resnik for the period of T−2:05 (prior to launch) through approximately T+73 seconds, when loss of all data occurred. The operational recorder was automatically activated at T−2:05; normally it runs throughout the entire mission. During the period of the prelaunch and the launch phase covered by the voice tape, Mission Specialist 3 Ronald E. McNair, Payload Specialist 1 S. Christa McAuliffe, and Payload Specialist 2 Gregory B. Jarvis were seated in the middeck and could monitor all voice activity but did not make any voice reports or comments.

Because of the concern over the loss of the seven astronauts, on 28 July 1986 Rear Admiral Richard H. Truly, NASA's associate administrator for Space Flight and a former astronaut, released the second document printed here—a report from Dr. Joseph P. Kerwin, biomedical specialist from the Johnson Space Center in Houston, Texas, relating to the astronauts' deaths.

The *Challenger* accident and the ensuing investigation invited comparison with events that followed the launch-pad fire of the *Apollo 204* spacecraft almost 19 years earlier. That fire had resulted in the deaths of three astronauts. During the earlier accident a politically strong administrator was at the helm of NASA; James E. Webb persuaded the White House to allow NASA to take the lead in the accident investigation. That investigation was largely technical, but it was sufficiently rigorous and critical to be credible. It resulted primarily in engineering changes; whatever managerial changes Webb made as a result were not extreme, lest the agency's entire management corps be cast into confusion. In contrast, after the *Challenger* accident NASA's internal investigation took a back seat to the work of a White House–appointed commission, chaired by former secretary of state William P. Rogers. NASA was unable to seize the initiative because, among other factors, its own top management was in disarray. The report of the "Rogers Commission" was deliberate and thorough. As the excerpt published here suggests, it gave as much emphasis to the accident's managerial origins as to its technical origins.

Document 17
TRANSCRIPT OF THE *CHALLENGER* CREW COMMENTS FROM THE OPERATIONAL RECORDER (28 JANUARY 1986)

CDR: Scobee
PLT: Smith
MS 1: Onizuka
MS 2: Resnik

(The references to "NASA" indicate explanatory references NASA provided to the Presidential Commission.)

(Min:Sec)	Position	Comment
T−2:05	MS 2	Would you give that back to me?
T−2:03	MS 2	Security blanket.
T−2:02	MS 2	Hmm.
T−1:58	CDR	Two minutes downstairs; you gotta watch running down there?
(NASA: Two minutes till launch.)		
T−1:47	PLT	OK there goes the lox arm.
(NASA: Liquid oxygen supply arm to ET.)		
T−1:46	CDR	Goes the beanie cap.
(NASA: Liquid oxygen vent cap.)		
T−1:44	MS 1	Doesn't it go the other way?
T−1:42	[Laughter]	
T−1:39	MS 1	Now I see it; I see it.
T−1:39	PLT	God I hope not Ellison.
T−1:38	MS 1	I couldn't see it moving; it was behind the center screen.
(NASA: Obstructed view of liquid oxygen supply arm.)		
T−1:33	MS 2	Got your harnesses locked?
(NASA: Seat restraints.)		
T−1:29	PLT	What for?
T−1:28	CDR	I won't lock mine; I might have to reach something.
T−1:24	PLT	Ooh kaaaay.
T−1:04	MS 1	Dick's thinking of somebody there.

(Min:Sec)	Position	Comment
T−1:03	CDR	Unhuh.
T−59	CDR	One minute downstairs.
(NASA: One minute till launch.)		
T−52	MS 2	Cabin pressure is probably going to give us an alarm.
(NASA: Caution and warning alarm. Routine occurrence during prelaunch.)		
T−50	CDR	OK.
T−47	CDR	OK there.
T−43	PLT	Alarm looks good.
(NASA: Cabin pressure is acceptable.)		
T−42	CDR	OK.
T−40	PLT	Ullage pressures are up.
(NASA: External tank ullage pressure.)		
T−34	PLT	Right engine helium tank is just a little bit low.
(NASA: SSME supply helium pressure.)		
T−32	CDR	It was yesterday, too.
T−31	PLT	OK.
T−30	CDR	Thirty seconds down there.
(NASA: 30 seconds till launch.)		
T−25	PLT	Remember the red button when you make a roll call.
(NASA: Precautionary reminder for communications configuration.)		
T−23	CDR	I won't do that; thanks a lot.
T−15	CDR	Fifteen.
(NASA: 15 seconds till launch.)		
T−6	CDR	There they go, guys.
(NASA: SSME ignition.)		
	MS 2	All right.
	CDR	Three at a hundred.
(NASA: SSME thrust level at 100% for all 3 engines.)		
T+0	MS 2	Aaall riiight.
T+1	PLT	Here we go.
(NASA: Vehicle motion.)		

(Min:Sec)	Position	Comment
T+7	CDR	Houston, *Challenger* roll program.

(NASA: Initiation of vehicle roll program.)

T+11	PLT	Go, you Mother.
T+14	MS 1	LVLH.

(NASA: Reminder for cockpit switch configuration change. Local vertical/local horizontal.)

T+15	MS 2	(Expletive) hot.
T+16	CDR	Ooohh-kaaay.
T+19	PLT	Looks like we've got a lotta wind here today.
T+20	CDR	Yeah.
T+22	CDR	It's a little hard to see out my window here.
T+28	PLT	There's ten thousand feet and Mach point five.

(NASA: Altitude and velocity report.)

T+30	[Garble]	
T+35	CDR	Point nine.

(NASA: Velocity report, 0.9 Mach.)

T+40	PLT	There's Mach one.

(NASA: Velocity report, 1.0 Mach.)

T+41	CDR	Going through nineteen thousand.

(NASA: Altitude report, 19,000 ft.)

T+43	CDR	OK, we're throttling down.

(NASA: Normal SSME thrust reduction during maximum dynamic pressure region.)

T+57	CDR	Throttling up.

(NASA: Throttle up to 104 percent after maximum dynamic pressure.)

T+58	PLT	Throttle up.
T+59	CDR	Roger.
T+60	PLT	Feel that mother go.
T+60		Woooohoooo.
T+1:02	PLT	Thirty-five thousand going through one point five.

(Min:Sec)	Position	Comment
(NASA: Altitude and velocity report, 35,000 ft., 1.5 Mach.)		
T+1:05	CDR	Reading four eighty six on mine.
(NASA: Routine airspeed indicator check.)		
T+1:07	PLT	Yep, that's what I've got, too.
T+1:10	CDR	Roger, go at throttle up.
(NASA: SSME at 104 percent.)		
T+1:13	PLT	Uhoh.
T+1:13	LOSS OF ALL DATA.	

Source: "Transcript of the *Challenger* Crew Comments from the Operational Recorder," 28 January 1986, NASA Historical Reference Collection, History Office, NASA Headquarters, Washington, DC.

Document 18
DR. KERWIN'S REPORT TO ADMIRAL TRULY
(28 JULY 1986)

RADM Richard H. Truly
Associate Administrator for Space Flight
NASA Headquarters
Code M
Washington, DC 20546

Dear Admiral Truly:

The search for wreckage of the *Challenger* crew cabin has been completed. A team of engineers and scientists has analyzed the wreckage and all other available evidence in an attempt to determine the cause of death of the *Challenger* crew. This letter is to report to you on the results of this effort. The findings are inconclusive. The impact of the crew compartment with the ocean surface was so violent that evidence of damage occurring in the seconds which followed the explosion was masked. Our final conclusions are:

the cause of death of the *Challenger* astronauts cannot be positively determined; the forces to which the crew were exposed during Orbiter breakup were probably not sufficient to cause death or serious injury; and the crew possibly, but not certainly, lost consciousness in the seconds following Orbiter breakup due to in-flight loss of crew module pressure.

Our inspection and analyses revealed certain facts which support the above conclusions, and these are related below: The forces on the Orbiter at breakup were probably too low to cause death or serious injury to the crew but were sufficient to separate the crew compartment from the forward fuselage, cargo bay, nose cone, and forward reaction control compartment. The forces applied to the Orbiter to cause such destruction clearly exceed its design limits. The data available to estimate the magnitude and direction of these forces included ground photographs and measurements from onboard accelerometers, which were lost two-tenths of a second after vehicle breakup.

Two independent assessments of these data produced very similar estimates. The largest acceleration pulse occurred as the Orbiter forward fuselage separated and was rapidly pushed away from the external tank. It then pitched nose-down and was decelerated rapidly by aerodynamic forces. There are uncertainties in our analysis; the actual breakup is not visible on photographs because the Orbiter was hidden by the gaseous cloud surrounding the external tank. The range of most probable maximum accelerations is from 12 to 20 G's in the vertical axis. These accelerations were quite brief. In two seconds, they were below four G's; in less than ten seconds, the crew compartment was essentially in free fall. Medical analysis indicates that these accelerations are survivable, and that the probability of major injury to crew members is low.

After vehicle breakup, the crew compartment continued its upward trajectory, peaking at an altitude of 65,000 feet approximately 25 seconds after breakup. It then descended, striking the ocean surface about two minutes and forty-five seconds after breakup at a velocity of about 207 miles per hour. The forces imposed by this impact approximated 200 G's, far in excess of the structural limits of the crew compartment or crew survivability levels.

The separation of the crew compartment deprived the crew of Orbiter-supplied oxygen, except for a few seconds supply in the lines. Each crew member's helmet was also connected to a personal egress air pack (PEAP) containing an emergency supply of breathing air (not oxygen) for ground egress emergencies, which must be manually activated to be available. Four PEAP's were recovered, and there is evidence that three had been activated. The nonactivated PEAP was identified as the Commander's, one of the others as the Pilot's, and the remaining ones could not be associated with any crew member. The evidence indicates that the PEAP's were not activated due to water impact.

It is possible, but not certain, that the crew lost consciousness due to an in-flight loss of crew module pressure. Data to support this is:

The accident happened at 48,000 feet, and the crew cabin was at that altitude or higher for almost a minute. At that altitude, without an oxygen supply, loss of cabin pressure would have caused rapid loss of consciousness and it would not have been regained before water impact. PEAP activation could have been an instinctive response to unexpected loss of cabin pressure. If a leak developed in the crew compartment as a result of structural damage during or after breakup (even if the PEAP's had been activated), the breathing air available would not have prevented rapid loss of consciousness. The crew seats and restraint harnesses showed patterns of failure, which demonstrates that all the seats were in place and occupied at water impact with all harnesses locked. This would likely be the case had rapid loss of consciousness occurred, but it does not constitute proof.

Much of our effort was expended attempting to determine whether a loss of cabin pressure occurred. We examined the wreckage carefully, including the crew module attach points to the fuselage, the crew seats, the pressure shell, the flight deck and middeck floors, and feedthroughs for electrical and plumbing connections. The windows were examined and fragments of glass analyzed chemically and microscopically. Some items of equipment stowed in lockers showed damage that might have occurred due to decompression; we experimentally decompressed similar items without conclusive results.

Impact damage to the windows was so extreme that the presence or absence of in-flight breakage could not be determined. The estimated breakup forces would not in themselves have broken the windows. A broken window due to flying debris remains a possibility; there was a piece of debris imbedded in the frame between two of the forward windows. We could not positively identify the origin of the debris or establish whether the event occurred in flight or at water impact. The same statement is true of the other crew compartment structure. Impact damage was so severe that no positive evidence for or against in-flight pressure loss could be found.

Finally, the skilled and dedicated efforts of the team from the Armed Forces Institute of Pathology, and their expert consultants, could not determine whether in-flight lack of oxygen occurred, nor could they determine the cause of death.

[signed]
Joseph P. Kerwin

Source: Joseph P. Kerwin, Johnson Space Center, to RADM Richard H. Truly, NASA Associate Administrator for Space Flight, NASA Historical Reference Collection, History Office, NASA Headquarters, Washington, DC.

Document 19
PRESSURES ON THE SYSTEM

With the 1982 completion of the orbital flight test series, NASA began a planned acceleration of the Space Shuttle launch schedule. One early plan contemplated an eventual rate of a mission a week, but realism forced several downward revisions. In 1985, NASA published a projection calling for an annual rate of 24 flights by 1990. Long before the *Challenger* accident, however, it was becoming obvious that even the modified goal of two flights a month was overambitious.

In establishing the schedule, NASA had not provided adequate resources for its attainment. As a result, the capabilities of the system were strained by the modest nine-mission rate of 1985, and the evidence suggests that NASA would not have been able to accomplish the 15 flights scheduled for 1986. These are the major conclusions of a Commission examination of the pressures and problems attendant upon the accelerated launch schedule.

On the same day that the initial orbital tests concluded—July 4, 1982—President Reagan announced a national policy to set the direction of the U.S. space program during the following decade. . . . From the inception of the Shuttle, NASA had been advertising a vehicle that would make space operations "routine and economical." The greater the annual number of flights, the greater the degree of routinization and economy, so heavy emphasis was placed on the schedule. However, the attempt to build up to 24 missions a year brought a number of difficulties, among them the compression of training schedules, the lack of spare parts, and the focusing of resources on near-term problems.

One effect of NASA's accelerated flight rate and the agency's determination to meet it was the dilution of the human and material resources that could be applied to any particular flight.

The part of the system responsible for turning the mission requirements and objectives into flight software, flight trajectory information and crew training materials was struggling to keep up with the flight rate in late 1985, and forecasts showed it would be unable to meet its milestones for 1986. It was falling behind because its resources were strained to the limit, strained by the flight rate itself and by the constant changes it was forced to respond to within that accelerating schedule. Compounding the problem was the fact that NASA had difficulty evolving from its single-flight focus to a system that could efficiently support the projected flight rate. It was slow in developing a hardware maintenance plan for its reusable fleet and slow in developing the capabilities that would allow it to handle the higher volume of work and training associated with the increased flight frequency.

Pressures developed because of the need to meet customer commitments, which translated into a requirement to launch a certain number of flights per year and to launch them on time. Such considerations may occasionally have obscured engineering concerns. Managers may have forgotten—partly because of past success, partly because of their own well-nurtured image of the program—that the Shuttle was still in a research and development phase. . . .

Planning of a Mission

The planning and preparation for a Space Shuttle flight require close coordination among those making the flight manifest, those designing the flight and the customers contracting NASA's services. The goals are to establish the manifest; define the objectives, constraints and capabilities of the mission; and translate those into hardware, software and flight procedures.

There are major program decision points in the development of every Shuttle flight. At each of these points, sometimes called freeze points, decisions are made that form the basis for further engineering and product development. The disciplines affected by these freeze points include integration hardware, engineering, crew timeline, flight design and crew training.

The first major freeze point is at launch minus 15 months. At that time the flight is officially defined: the launch date, Orbiter and major payloads are all specified, and initial design and engineering are begun based on this information.

The second major freeze point is at launch minus 7.7 months, the cargo integration hardware design. Orbiter vehicle configuration, flight design and software requirements are agreed to and specified. Further design and engineering can then proceed.

Another major freeze point is the flight planning and stowage review at launch minus five months. At that time, the crew activity timeline and the crew compartment configuration, which includes middeck payloads and payload specialist assignments, are established. Final design, engineering and training are based on these products. . . .

Changes in the Manifest

Each process in the production cycle is based on information agreed upon at one of the freeze points. If that information is later changed, the process may have to be repeated. The change could be a change in manifest or a change to the Orbiter hardware or software. The hardware and software changes in 1985 usually were mandatory changes; perhaps some of the manifest changes were not.

The changes in the manifest were caused by factors that fall into four general categories: hardware problems, customer requests, operational constraints and external factors. . . . The following examples illustrate that a single proposed change can have extensive impact, not because the change itself is particularly difficult to accommodate (though it may be), but because each change necessitates four or five other changes. The cumulative effect can be substantial. . . .

When a change occurs, the program must choose a response and accept the consequences of that response. The options are usually either to maximize the benefit to the customer or to minimize the adverse impact on Space Shuttle operations. If the first option is selected, the consequences will include short-term and/or long-term effects. . . .

"Operational" Capabilities

For a long time during Shuttle development, the program focused on a single flight, the first Space Shuttle mission. When the program became "operational," flights came more frequently, and the same resources that had been applied to one flight had to be applied to several flights concurrently. Accomplishing the more pressing immediate requirements diverted attention from what was happening to the system as a whole. That appears to be one of the many telling differences between a "research and development" program and an "operational program." Some of the differences are philosophical, some are attitudinal and some are practical.

Elements within the Shuttle program tried to adapt their philosophy, their attitude and their requirements to the "operational era." But that era came suddenly, and in some cases, there had not been enough preparation for what "operational" might entail. For example, routine and regular post-flight maintenance and inspections are critical in an operational program; spare parts are critical to flight readiness in an operational fleet; and the software tools and training facilities developed during a test program may not be suitable for the high volume of work required in an operational environment. In many respects, the system was not prepared to meet an "operational" schedule.

As the Space Shuttle system matured, with numerous changes and compromises, a comprehensive set of requirements was developed to ensure the success of a mission. What evolved was a system in which the preflight processing, flight planning, flight control and flight training were accomplished with extreme care applied to every detail. This process checked and rechecked everything, and though it was both labor- and time-intensive, it was appropriate and necessary for a system still in the developmental phase. This process, however, was not capable of meeting the flight rate goals.

After the first series of flights, the system developed plans to accomplish what was required to support the flight rate. The challenge was to streamline the process through automation, standardization, and centralized management, and to convert from the developmental phase to the mature system without a compromise in quality. It required that experts carefully analyze their areas to determine what could be standardized and automated, then take the time to do it.

But the increasing flight rate had priority—quality products had to be ready on time. Further, schedules and budgets for developing the needed facility improvements were not adequate. Only the time and resources left after supporting the flight schedule could be directed toward efforts to streamline and standardize. In 1985, NASA was attempting to develop the capabilities of a production system. But it was forced to do that while responding—with the same personnel—to a higher flight rate.

At the same time the flight rate was increasing, a variety of factors reduced the number of skilled personnel available to deal with it. These included retirements, hiring freezes, transfers to other programs like the Space Station and transitioning to a single contractor for operations support.

The flight rate did not appear to be based on assessment of available resources and capabilities and was not reduced to accommodate the capacity of the work force. For example, on January 1, 1986, a new contract took effect at Johnson that consolidated the entire contractor work force under a single company. This transition was another disturbance at a time when the work force needed to be performing at full capacity to meet the 1986 flight rate. In some important areas, a significant fraction of workers elected not to change contractors. This reduced the work force and its capabilities, and necessitated intensive training programs to qualify the new personnel. According to projections, the work force would not have been back to full capacity until the summer of 1986. This drain on a critical part of the system came just as NASA was beginning the most challenging phase of its flight schedule. . . .

Responding to Challenges and Changes

Another obstacle in the path toward accommodation of a higher flight rate is NASA's legendary "can-do" attitude. The attitude that enabled the agency to put men on the moon and to build the Space Shuttle will not allow it to pass up an exciting challenge—even though accepting the challenge may drain resources from the more mundane (but necessary) aspects of the program.

A recent example is NASA's decision to perform a spectacular retrieval of two communications satellites whose upper-stage motors had failed to raise them to the proper geosynchronous orbit. NASA itself then proposed to the insurance companies who owned the failed satellites that the agency design a mission to rendezvous with them in turn and that an astronaut in a jet backpack fly over to escort the satellites into the Shuttle's payload bay for a return to Earth.

The mission generated considerable excitement within NASA and required a substantial effort to develop the necessary techniques, hardware and procedures. The mission was conceived, created, designed and accomplished within 10 months. The result, mission 51-A (November 1984), was a resounding success, as both failed satellites were successfully returned to Earth. The retrieval mission vividly demonstrated the service that astronauts and the Space Shuttle can perform. . . .

The satellite retrieval missions were not isolated occurrences. Extraordinary efforts on NASA's part in developing and accomplishing missions will, and should, continue, but such efforts will be a substantial additional drain on resources. NASA cannot both accept the relatively spur-of-the-moment missions that its "can-do" attitude tends to generate and also maintain the planning and scheduling discipline required to operate as a "space truck" on a routine and cost-effective basis. As the flight rate increases, the cost in resources and the accompanying impact on future operations must be considered when infrequent but extraordinary efforts are undertaken. The system is still not sufficiently developed as a "production line" process in terms of planning or implementation procedures. It cannot routinely or even periodically accept major disruptions without considerable cost. NASA's attitude historically has reflected the position that "We can do anything," and while that may essentially be true, NASA's optimism must be tempered by the realization that it cannot do everything. . . .

It is important to determine how many flights can be accommodated, and accommodated safely. NASA must establish a realistic level of expectation, then approach it carefully. Mission schedules should be based on a realistic assessment of what NASA can do safely and well, not on what is possible with maximum effort. The ground rules must be established firmly, and then enforced. . . .

Findings

1. The capabilities of the system were stretched to the limit to support the flight rate in winter 1985/1986. Projections into the spring and summer of 1986 showed a clear trend; the system, as it existed, would have been unable to deliver crew training software for scheduled flights by the designated dates. The result would have been an unacceptable compression of the time available for the crews to accomplish their required training.

2. Spare parts are in critically short supply. The Shuttle program made a conscious decision to postpone spare parts procurements in favor of budget items of perceived higher priority. Lack of spare parts would likely have limited flight operations in 1986.

3. Stated manifesting policies are not enforced. Numerous late manifest

changes (after the cargo integration review) have been made to both major payloads and minor payloads throughout the Shuttle program. . . .

4. The scheduled flight rate did not accurately reflect the capabilities and resources. . . .

5. Training simulators may be the limiting factor on the flight rate: the two current simulators cannot train crews for more than 12–15 flights per year.

6. When flights come in rapid succession, current requirements do not ensure that critical anomalies occurring during one flight are identified and addressed appropriately before the next flight.

Source: *Report of the Presidential Commission on the Space Shuttle* Challenger *Accident*, Vol. I (Washington, DC: Government Printing Office, 1986), pp. 164–177.

A PROMISE OF SPACE EXPLORATION FOR THE NEW CENTURY

As the century comes to a close, several people in the United States have pressed for an aggressive effort to reach Mars, still one of the most enticing planets in the solar system. Robert Zubrin, a leader of the National Space Society, is one of the most persistent of these advocates. His essay speaks to both the romance and the necessity of space exploration at the dawn of a new millennium.

Document 20
A NEW MARTIAN FRONTIER: RECAPTURING THE SOUL OF AMERICA

A bit more than 100 years ago, a young professor of history from the relatively obscure University of Wisconsin got up to speak at the annual conference of the American Historical Association. Frederick Jackson Turner's talk was the last one in the evening session. A series of excruciatingly boring papers on topics so obscure that kindness forbids even reprinting their titles preceded Turner's address, yet the majority of the conference participants stayed to hear him.

Perhaps a rumor had gotten afoot that something important was about to be said. If so, it was correct, for in one bold sweep of brilliant insight Turner laid bare the source of the American soul. It was not legal theories, precedents, traditions, national or racial stock that was the source of the egalitarian democracy, individualism and spirit of innovation that characterized America. It was the existence of the frontier.

"To the frontier the American intellect owes its striking characteristics," Turner roared. "That coarseness of strength combined with acuteness and inquisitiveness; that practical, inventive turn of mind, quick to find expedients; that masterful grasp of material things, lacking in the artistic but powerful to effect great ends; that restless, nervous energy; that dominant individualism, working for good and evil, and withal that buoyancy and exuberance that comes from freedom—these are the traits of the frontier, or traits called out elsewhere because of the existence of the frontier."

Turner rolled on, entrancing his audience, "For a moment, at the frontier, the bonds of custom are broken and unrestrained triumphant. There is no tabula rasa. The stubborn American environment is there with its imperious summons to accept its conditions; the inherited ways of doing things are also there; and yet, in spite of the environment, and in spite of custom, each frontier did indeed furnish a new opportunity, a gate of escape from the bondage of the past; and freshness, and confidence, and scorn of older society, impatience of its restraints and its ideas, and indifference to its lessons, have accompanied the frontier."

The Turner thesis was a bombshell. Within a few years an entire school of historians proceeded to demonstrate that not only American culture, but the entire Western progressive humanist civilization that America has generally represented resulted from the Great Frontier of global settlement opened to Europe by the Age of Exploration.

Turner presented his paper in 1893. Just three years earlier, in 1890, the American frontier had been declared closed: the line of settlement that had always defined the furthermost existence of western expansion had actually met the line of settlement coming east from California. Now, a century later, we face a question that has grown over the course of the past 100 years—what if the frontier is gone? What happens to America and all it has stood for? Can a free, egalitarian, democratic, innovating society with a can-do spirit be preserved in the absence of room to grow?

We see around us now an ever more apparent loss of vigor of American society: increasing fixity of the power structure and bureaucratization of all levels of society; impotence of political institutions to carry off great projects; the cancerous proliferation of regulations affecting all aspects of public, private and commercial life; the spread of irrationalism; the banalization of popular culture; the loss of willingness by individuals to take risks, to fend or think for themselves; economic stagnation and decline; the deceleration of the rate of technological innovation and a loss of belief in the idea of progress itself. Everywhere you look, the writing is on the wall.

Without a frontier from which to breathe life, the spirit that gave rise to the progressive humanistic culture that America has offered to the world for the past several centuries is fading. The issue is not just one of national loss—human progress needs a vanguard, and no replacement is in sight.

The creation of a new frontier thus presents itself as America's and humanity's greatest social need. Nothing is more important: Apply what palliatives you will, without a frontier to grow in, not only American society, but the entire global civilization based upon Western enlightenment values of humanism, reason, science and progress will ultimately die.

I believe that humanity's new frontier can only be on Mars. Why Mars? Why not on Earth, under the oceans or in such remote regions as Antarctica? And if it must be in space, why on Mars? Why not on the Moon or in artificial satellites in orbit about the Earth?

It is true that settlements on or under the sea or in Antarctica are entirely possible, and their establishment and access would be much easier than that of Martian colonies. Nevertheless, the fact of the matter is that at this point in history such terrestrial developments cannot meet an essential requirement for a frontier—to wit, they are insufficiently remote to allow for the free development of a new society. In this day and age, with modern terrestrial communication and transportation systems, no matter how remote or hostile the spot on Earth, the cops are too close. If people are to have the dignity that comes with making their own world, they must be free of the old.

Why then, not the Moon? The answer is because there's not enough there. True, the Moon has a copious supply of most metals and oxygen, in the form of oxidized rock, and a fair supply of solar energy, but that's about it. For all intents and purposes, the Moon has no hydrogen, nitrogen, or carbon, and these are three of the four elements most necessary for life. (They're present in the lunar soil in parts per million quantities, somewhat like gold in sea water. If there were concrete on the Moon, lunar colonists would mine it to get its water out.) You could bring seeds to the Moon and grow plants in enclosed greenhouses, but nearly every atom of carbon, nitrogen, and hydrogen that goes into making those plants would have to be imported from another planet. While sustaining a lunar scientific base under such conditions is relatively straightforward, growing a civilization there would be impossible. The difficulties in supporting significant populations in artificial orbiting space colonies would be even greater.

Mars has what it takes. It's far enough away to free its colonists from intellectual, legal, or cultural domination by the old world, and rich enough in resources to give birth to the new. The Red Planet may appear at first glance to be a desert, but beneath its sands are oceans of water in the form of permafrost, enough in fact, if it were melted and Mars' terrain were smoothed out, to cover the entire planet with an ocean several hundred meters deep. Mars' atmosphere is mostly carbon dioxide, providing enormous supplies of the two most important biological elements in a chemical form from which they can be directly taken up and incorporated into plant life. Mars has nitrogen too, both as a minority constituent (3%) in its atmosphere and probably as nitrate beds in its soil as well. For the rest, all the metals, silicon, sulfur, phosphorus, inert gases, and other raw materials needed to create not only life but an advanced technological civilization, can readily be found on Mars.

Mars is remote and can be settled. Indeed, one can imagine an engineering capability established on the planet within a hundred years of the first outposts that would allow for the transformation of the planet to the warm, wet conditions of Mars' primitive past, making a desert world into a new home for a new spectrum of descendants of terrestrial life. The fact that Mars can be settled and altered defines it as the New World that can create the basis for a positive future for terrestrial humanity for the next several centuries.

Why Humanity Needs Mars

To see best why twenty-first-century humanity will desperately need an open frontier on Mars, we need to look at modern Western humanist culture and see what makes it so much more desirable a mode of society than anything that has ever existed before. Then we need to see how everything we hold dear will be wiped out if the frontier remains closed.

The essence of humanist society is that it values human beings—human life and human rights are held precious beyond price. Such notions have been for several thousand years the core philosophical values of Western civilization, dating back to the Greeks and the Judeo-Christian ideas of the divine nature of the human spirit. Yet these values could never be implemented as a practical basis for the organization of society until the great explorers of the age of discovery threw open a New World in which the dormant seed of medieval Christendom could grow and blossom forth into something the likes of which the world had never seen before.

The problem with medieval Christendom was that it was fixed—it was a play for which the script had been written and the leading roles both chosen and assigned. The problem was not that there were insufficient natural resources to go around—medieval Europe was not heavily populated, there were plenty of forests and other wild areas—the problem was that all the resources were owned. A ruling class had been selected and a set of ruling institutions, ideas and customs had been selected, and by the law of "Survival of the Firstest," none of these could be displaced. Furthermore, not only the leading roles had been chosen, but also those of the supporting cast and chorus, and there were only so many such parts to go around. If you wanted to keep your part, you had to keep your place, and there was no place for someone without a place.

The New World changed all that by supplying a place in which there were no established ruling institutions, an improvisational theater big enough to welcome all comers with no parts assigned. On such a stage, the players are not limited to the conventional role of actors, they become playwrights and directors as well. The unleashing of creative talent that such a novel situation allows is not only a great deal of fun for those lucky enough to be involved, it changes the view of the spectators as to the capabilities of actors in general. People who had no role in the old society could define their role in the new. People who did not "fit in" in the old world could discover and demonstrate that far from being worthless, they were invaluable in the new, whether they went there or not.

The New World destroyed the basis of aristocracy and created the basis of democracy. It allowed the development of diversity by allowing escape from those institutions that imposed uniformity. It destroyed a closed intellectual world by importing unsanctioned data and experience. It allowed progress by escaping the hold of those institutions whose continued rule required continued stagnation, and it drove progress by defining a situation in which innovation to maximize the capabilities of the limited population available was desperately needed. It raised the dignity of workers by raising the price of labor and by demonstrating for all to see that human beings can be the creators of their world, and not just its inhabitants. (In America, during the nineteenth century when cities were rapidly being built, there were people who understood that America was not something one simply lived in—it was a place one helped make. People were not simply inhabitants of the world. They were makers of the world.)

Now consider the probable fate of humanity in the twenty-first century under two conditions—with a Martian frontier and without it.

In the twenty-first century, without a Martian frontier, there is no question that human diversity will decline severely. Already, in the late twentieth century, advanced communication and transportation technologies have eroded the healthy diversity of human cultures on Earth, and this tendency can only accelerate in the twenty-first. On the other hand, if the Martian frontier is opened, then this same process of technological advance will also enable us to establish a new branch of human culture on Mars and eventually worlds beyond. The precious diversity of humanity can thus be preserved on a broader field, but only on a broader field. One world will be just too small a domain to allow the preservation of the diversity needed not just to keep life interesting, but to assure the survival of the human race.

Technological Innovation

Without the opening of a new frontier on Mars, continued Western civilization faces the risk of technological stagnation. To some this may appear to be an outrageous statement, as the present age is frequently cited as one of technological wonders. In fact, however, the rate of progress within our society has been decreasing, and at an alarming rate. To see this, it is only necessary to step back and compare the changes that have occurred in the past 30 years with those that occurred in the two preceding 30-year periods.

Between 1903 and 1933 the world was revolutionized: Cities were electrified; telephones and broadcast radio became common; talking motion pictures appeared; automobiles became practical; and aviation progressed from the Wright Flyer to the DC-3 and Hawker Hurricane. Between 1933 and 1963 the world changed again, with the introduction of color television, communication satellites and interplanetary spacecraft, computers, antibiotics, scuba gear, nuclear power, Atlas, Titan, and Saturn rockets, Boeing 727's and SR-71's.

Compared to these changes, the technological innovations from 1963 to the present are insignificant. Immense changes should have occurred during this period, but did not. Had we been following the previous 60 years' technological trajectory, we today would have videotelephones, solar powered cars, maglev trains, fusion reactors, hypersonic intercontinental travel, regular passenger transportation to orbit, undersea cities, open-sea mariculture and human settlements on the Moon and Mars.

Consider a nascent Martian civilization: Its future will depend critically upon the progress of science and technology. Just as the inventions produced by the "Yankee Ingenuity" of frontier America were a powerful driving force on world-wide human progress in the nineteenth century, so the "Martian Ingenuity" born in a culture that puts the utmost premium on intelligence, practical education and the determination required to make real contributions will provide much more than its fair share of the scientific and technological breakthroughs that will dramatically advance the human condition in the twenty-first century.

A prime example of the Martian frontier driving new technology will undoubtedly be found in the arena of energy production. As on Earth, a copious supply of energy will be crucial to the success of Mars settlements. The Red Planet does have one major energy resource that we currently know about: deuterium, which can be used as the fuel in nearly waste-free thermonuclear fusion reactors. Earth has large amounts of deuterium too, but with all of its existing investments in other, more polluting forms of energy production, the research that would make possible practical fusion power reactors has been allowed to stagnate. The Martian colonists are certain to be much more determined to get fusion on-line, and in doing so will massively benefit the mother planet as well.

The parallel between the Martian frontier and that of nineteenth-century America as technology drivers is, if anything, vastly understated. America drove technological progress in the last century because its western frontier created a perpetual labor shortage in the east, thus forcing the development of labor-saving machinery and providing a strong incentive for improvement of public education so that the skills of the limited labor force available could be maximized. This condition no longer holds true in America. In fact, far from prizing each additional citizen, immigrants are no longer welcome here and a vast "service sector" of bureaucrats and menials has been created to absorb the energies of the majority of the population which is excluded from the productive parts of the economy. Thus in the late twentieth century, and increasingly in the twenty-first, each additional citizen is and will be regarded as a burden.

On twenty-first-century Mars, on the other hand, conditions of labor shortage will apply with a vengeance. Indeed, it can be safely said that no commodity on twenty-first-century Mars will be more precious, more highly valued and more dearly paid for than human labor time. Workers on Mars will be paid more and treated better than their counterparts on Earth. Just as the example of nineteenth-century America changed the way the common man was regarded and treated in Europe, so the impact of progressive Martian social conditions will be felt on Earth as well as on Mars. A new standard will be set for a higher form of humanist civilization on Mars, and, viewing it from afar, the citizens of Earth will rightly demand nothing less for themselves.

Politics on Earth with Humans on Mars

The frontier drove the development of democracy in America by creating a self-reliant population which insisted on the right to self-government. It is doubtful that democracy can persist without such people. True, the trappings of democracy exist in abundance in America today, but meaningful public participation in the process has all but disappeared. Consider that no representative of a new political party has been elected President of the Unites States since 1860. Likewise, neighborhood political clubs and ward structures that once allowed citizen participation in party deliberations have vanished. And with re-election rates typically close to 95 percent, the U.S. Congress is hardly susceptible to the people's will. Regardless of the will of Congress, the real laws, covering ever broader areas of economic and social life, are increasingly being made by a plethora of regulatory agencies whose officials do not even pretend to have been elected by anyone.

Democracy in America and elsewhere in Western civilization needs a shot in the arm. That boost can only come from the example of a frontier people whose civilization incorporates the ethos that breathed the spirit into democracy in America in the first place. As Americans showed Europe in the last century, so in the next the Martians can show us the way away from oligarchy.

There are greater threats that a humanist society faces in a closed work than the return of oligarchy, and if the frontier remains closed we are certain to face them in the twenty-first century. These threats are the spread of various sorts of anti-human ideologies and the development of political institutions that incorporate the notions that spring from them as a basis of operation. At the top of the list of such pathological ideas that tend to spread naturally in a closed society is the Malthus theory, which holds that since the world's resources are more or less fixed, population growth must be restricted or all of us will descend into bottomless misery.

Malthusianism is scientifically bankrupt—all predictions made upon it have been wrong, because human beings are not mere consumers of resources. Rather, we create resources by the development of new technologies that find use for new raw materials. The more people, the faster the rate of innovation. This is why (contrary to Malthus) as the world's population has increased, the standard of living has increased, and at an accelerating rate. Nevertheless, in a closed society Malthusianism has the appearance of self-evident truth, and herein lies the danger. It is not enough to argue against Malthusianism in the abstract—such debates are not settled in academic journals. Unless people can see broad vistas of unused resources in front of them, the belief in limited resources tends to follow as a matter of course. If the idea is accepted that the world's resources are fixed, then each person is ultimately the enemy of every other person, and each race or nation is the enemy of every other race or nation. Only in a universe of unlimited resources can all men be brothers.

Mars Beckons

Western humanist civilization as we know and value it today was born in expansion, grew in expansion, and can only exist in a dynamic expansion. While some form of human society might persist in a non-expanding world, that society will not feature freedom, creativity, individuality, or progress, and placing no value on those aspects of humanity that differentiate us from animals, it will place no value on human rights or human life as well.

Such a dismal future might seem an outrageous prediction, except for the fact that for nearly all of its history most of humanity has been forced to endure static modes of social organization, and the experience has not been a happy one. Free societies are the exception in human history—they have only existed during the four centuries of frontier expansion of the West. That history is now over. The frontier opened by the voyage of Christopher Columbus is now closed. If the era of Western humanist society is not to be seen by future historians as some kind of transitory golden age, a brief shining moment in an otherwise endless chronicle of human misery, then a new frontier must be opened. Mars beckons.

But Mars is only one planet, and with humanity's power over nature rising exponentially as they would in an age of progress that an open Martian frontier portends, the job of transforming and settling it is unlikely to occupy our energies for more than three or four centuries. Does the settling of Mars then simply represent an opportunity to "prolong but not save a civilization based upon dynamism"? Isn't it the case that humanist civilization is ultimately doomed anyway? I think not.

The universe is vast. Its resources, if we can access them, are truly infinite. During the four centuries of the open frontier on Earth, science and technology have advanced at an astonishing pace. The technological capabilities achieved during the twentieth century would dwarf the expectations of any observer from the nineteenth, seem like dreams to one from the eighteenth, and appear outright magical to someone from the seventeenth century. If the past four centuries of progress have multiplied our reach by so great a ratio, might not four more centuries of freedom do the same again? There is ample reason to believe that they would.

Terraforming Mars will drive the development of new and more powerful sources of energy; settling the Red Planet will drive the development of ever faster modes of space transportation. Both of these capabilities in turn will open up new frontiers ever deeper into the outer solar system, and the harder challenges posed by these new environments will drive the two key technologies of power and propulsion ever more forcefully, opening the path to the stars. The key is not to let the process stop. If it is allowed to stop for any length of time, society will crystallize into a static form that is inimical to the resumption of progress. That is what defines the present age as one of crisis. Our old frontier is closed, the first signs of social crystallization are clearly visible. Yet, progress, while slowing, is still extant; our people still believe in it and our ruling institutions are not yet incompatible with it.

We still possess the greatest gift of the inheritance of a 400-year-long Renaissance: to wit, the capacity to initiate another by opening the Martian frontier. If we fail to do so, our culture will not have that capacity long. Mars is harsh. Its settlers will need not only technology, but the scientific outlook, creativity and free-thinking individualistic inventiveness that stand behind it. Mars will not allow itself to be settled by people from a static society—those people won't have what it takes. We still do. Mars today waits for the children of the old frontier, but Mars will not wait forever.

Source: Robert Zubrin is a member of the Board of Directors of the National Space Society, Washington, DC. This document was originally published on the NASA Ames Research Center World Wide Web site at URL: http://cmex-www.arc.nasa.gov/MarsNews/Zubrin.html

Appendix

American Space Flights, 1961–1997

Spacecraft	Launch Date	Crew	Flight Time (days:hrs:mins)	Highlights
Mercury-Redstone 3	May 5, 1961	Alan B. Shepard	0:0:15	First U.S. flight; suborbital.
Mercury-Redstone 4	Jul. 21, 1961	Virgil I. Grissom	0:0:16	Suborbital; capsule sank after landing; astronaut safe.
Mercury-Atlas 6	Feb. 20, 1962	John H. Glenn, Jr.	0:4:55	First American to orbit.
Mercury-Atlas 7	May 24, 1962	M. Scott Carpenter	0:4:56	Landed 400 km beyond target.
Mercury-Atlas 8	Oct. 3, 1962	Walter M. Schirra	0:9:13	Landed 8 km from target.
Mercury-Atlas 9	May 15, 1963	L. Gordon Cooper	1:10:20	First U.S. flight exceeding 24 hours.
Gemini 3	Mar. 23, 1965	Virgil I. Grissom John W. Young	0:4:53	First U.S. 2-person flight; first manual maneuvers in orbit.
Gemini 4	Jun. 3, 1965	James A. McDivitt Edward H. White	4:1:56	21-min extravehicular activity (White).
Gemini 5	Aug. 21, 1965	L. Gordon Cooper Charles Conrad	7:22:55	Longest-duration human flight to date.
Gemini 7	Dec. 4, 1965	Frank Borman James A. Lovell	13:18:35	Longest human flight to date.
Gemini 6-A	Dec. 15, 1965	Walter M. Schirra Thomas P. Stafford	1:1:51	Rendezvous within 30 cm of Gemini 7.
Gemini 8	Mar. 16, 1966	Neil A. Armstrong David R. Scott	0:10:41	First docking of 2 orbiting spacecraft (Gemini 8 with an Agena target vehicle).
Gemini 9-A	Jun. 3, 1966	Thomas P. Stafford Eugene A. Cernan	3:0:21	Extravehicular activity; rendezvous.
Gemini 10	Jul. 18, 1966	John W. Young Michael Collins	2:22:47	First dual rendezvous (Gemini 10 with Agena 10 and Agena 8).
Gemini 11	Sep. 12, 1966	Charles Conrad, Jr. Richard F. Gordon, Jr.	2:23:17	First initial-orbit docking; first tethered flight; highest Earth-orbit altitude (1,372 km).
Gemini 12	Nov. 11, 1966	James A. Lovell, Jr. Edwin E. Aldrin, Jr.	3:22:35	Longest extravehicular activity to date (Aldrin, 5 hrs 37 min).

Apollo 7	Oct. 11, 1968	Walter M. Schirra, Jr. Donn F. Eisele R. Walter Cunningham	10:20:9	First U.S. 3-person mission.
Apollo 8	Dec. 21, 1968	Frank Borman James A. Lovell, Jr. William A. Anders	6:3:1	First human orbit(s) of Moon; Moon; first human departure of Earth's sphere of influence; highest speed attained in human flight to date.
Apollo 9	Mar. 3, 1969	James A. McDivitt David R. Scott Russell L. Schweickart	10:1:1	Successfully simulated in Earth orbit operation of lunar module to landing and takeoff from lunar surface and rejoining with command module.
Apollo 10	May 18, 1969	Thomas P. Stafford John W. Young Eugene A. Cernan	8:0:3	Successfully demonstrated complete system including lunar module to 14,300 m from the lunar surface.
Apollo 11	Jul. 16, 1969	Neil A. Armstrong Michael Collins Edwin E. Aldrin, Jr.	8:3:9	First human landing on Moon.
Apollo 12	Nov. 14, 1969	Charles Conrad, Jr. Richard F. Gordon, Jr. Alan L. Bean	10:4:36	Second human lunar landing. Explored surface of Moon, retrieved parts of Surveyor 3 spacecraft, which landed in Ocean of Storms on Apr. 19, 1967.
Apollo 13	Apr. 11, 1970	James A. Lovell Fred W. Haise, Jr. John L. Swigert, Jr.	5:22:55	Mission aborted; explosion in service module. Ship circled Moon, with crew using LM as "lifeboat" until just before reentry.
Apollo 14	Jan. 31, 1971	Alan B. Shepard Stuart A. Roosa Edgar D. Mitchell	9:0:2	Third human lunar landing. Mission demonstrated pinpoint landing capability and continued human exploration.
Apollo 15	Jul. 26, 1971	David R. Scott Alfred M. Worden James B. Irwin	12:7:12	Fourth human lunar landing and first Apollo "J" series mission, which carried Lunar Roving Vehicle. Worden's in-flight EVA of 38 min 12 sec was performed during return trip.
Apollo 16	Apr. 16, 1972	John W. Young Charles M. Duke, Jr. Thomas K. Mattingly II	11:1:51	Fifth human lunar landing, and second with Lunar Roving Vehicle.

Apollo 17	Dec. 7, 1972	Eugene A. Cernan Harrison H. Schmitt Ronald E. Evans	12:13:52	Sixth and final Apollo human lunar landing, again with roving vehicle.
Skylab 2	May 25, 1973	Charles Conrad, Jr. Joseph P. Kerwin Paul J. Weitz	28:0:50	Docked with Skylab 1 (launched uncrewed May 14) for 28 days. Repaired damaged station.
Skylab 3	Jul. 28, 1973	Alan L. Bean Jack R. Lousma Owen K. Garriott	59:11:9	Docked with Skylab 1 for more than 59 days.
Skylab 4	Nov. 16, 1973	Gerald P. Carr Edward G. Gibson William R. Pogue	84:1:16	Docked with Skylab 1 in long-duration mission; last of the Skylab program.
Apollo (ASTP)	Jul. 15, 1975	Thomas P. Stafford Donald K. Slayton Vance D. Brand	9:1:28	Docked with Soyuz 19 in joint experiments of ASTP mission.
Space Shuttle Columbia (Space Transportation System (STS)-1)	Apr. 12, 1981	John W. Young Robert L. Crippen	2:6:21	First flight of Space Shuttle Columbia, tested spacecraft in orbit. First landing of airplane-like craft from orbit for reuse.
Space Shuttle Columbia (STS-2)	Nov. 12, 1981	Joe H. Engle Richard H. Truly	2:6:13	Second flight of Space Shuttle Columbia, first scientific payload. Tested remote manipulator arm. Returned for reuse.
Space Shuttle Columbia (STS-3)	Mar. 22, 1982	Jack R. Lousma C. Gordon Fullerton	8:4:49	Third flight of Space Shuttle Columbia, second scientific payload (0SS 1). Second test of remote manipulator arm. Flight extended 1 day because of flooding at primary landing site; alternate landing site used. Returned for reuse.
Space Shuttle Columbia (STS-4)	Jun. 27, 1982	Thomas K. Mattingly II Henry W. Hartsfield, Jr.	7:1:9	Fourth flight of STS, first Department of Defense (DOD) payload, additional scientific payloads. Returned Jul. 4. Completed testing program. Returned for reuse.

Space Shuttle Columbia (STS-5)	Nov. 11, 1982	Vance D. Brand Robert F. Overmyer Joseph P. Allen William B. Lenoir	5:2:14	Fifth flight of STS. First operational flight. Launched two commercial satellites (SBS 3 and Anik C-3). First flight with 4 crew members. EVA test canceled when spacesuits malfunctioned.
Space Shuttle Challenger (STS-6)	Apr. 4, 1983	Paul J. Weitz Karol J. Bobko Donald H. Peterson Story Musgrave	5:0:24	Sixth flight of STS, launched TDRS 1.
Space Shuttle Challenger (STS-7)	Jun. 18, 1983	Robert L. Crippen Frederick H. Hauck John M. Fabian Sally K. Ride Norman T. Thagard	6:2:24	Seventh flight of STS, launched 2 commercial satellites (Anik C-2 and Palapa B-1) and retrieved SPAS 01. First flight with 5 crew members, including first woman U.S. astronaut.
Space Shuttle Challenger (STS-8)	Aug. 30, 1983	Richard H. Truly Daniel C. Brandenstein Dale A. Gardner Guion S. Bluford, Jr. William E. Thornton	6:1:9	Eighth flight of STS. Launched one commercial satellite (Insat l-B). First flight of U.S. black astronaut.
Space Shuttle Columbia (STS-9)	Nov. 28, 1983	John W. Young Brewster W. Shaw Owen K. Garriott Robert A. R. Parker Byron K. Lichtenberg Ulf Merbold	10:7:47	Ninth flight of STS. First flight of Spacelab 1. First flight of 6 crew members. First non-U.S. astronauts, one of whom was West German, to fly in U.S. space program (Merbold).
Space Shuttle Challenger (STS-41B)	Feb. 3, 1984	Vance D. Brand Robert L. Gibson Ronald E. McNair Roben L. Stewart	7:23:16	Tenth flight of STS; two communication satellites failed to achieve orbit. First use of Manned Maneuvering Unit (MMU) in space.
Space Shuttle Challenger (STS-41C)	Apr. 6, 1984	Robert L. Crippen Frances R. Scobee Terry J. Hart	6:23:41	Eleventh flight of STS. Deployment of Long-Duration Exposure Facility (LDEF-l), for later retrieval. Solar Maximum Satellite retrieved,

		George D. Nelson James D. van Hoften		repaired, and redeployed.
Space Shuttle Discovery (STS-41D)	Aug. 30, 1984	Henry Hartsfield Michael L. Coats Richard M. Mullane Steven A. Hawley Judith A. Resnick Charles D. Walker	6:0:56	Twelfth flight of STS. First flight of U.S. non-astronaut.
Space Shuttle Challenger (STS-41G)	Oct. 5, 1984	Robert L. Crippen Jon A. McBride Kathryn D. Sullivan Sally K. Ride David Leestma Paul D. Scully-Power Marc Garneau	8:5:24	Thirteenth flight of STS. First with 7 crew members, including first flight of two U.S. women and one Canadian (Garneau).
Space Shuttle Discovery (STS-51A)	Nov. 8, 1984	Frederick H. Hauck David M. Walker Joseph P. Allen Anna L. Fisher Dale A. Gardner	7:23:45	Fourteenth flight of STS. First retrieval and return of two disabled communications satellites (Westar 6, Palapa B2) to Earth.
Space Shuttle Discovery (STS-51C)	Jan. 24, 1985	Thomas K. Mattingly Loren J. Shriver Ellison S. Onizuka James F. Buchli Gary E. Payton	3:1:33	Fifteenth STS flight. Dedicated DOD mission.
Space Shuttle Discovery (STS-51D)	Apr. 12, 1985	Karol J. Bobko Donald E. Williams M. Rhea Seddon S. David Griggs Jeffrey A. Hoffman Charles D. Walker E. J. Garn	6:23:55	Sixteenth STS flight. Two communications satellites. First U.S. Senator in space (Garn).

Space Shuttle Challenger (STS-51B)	Apr. 29, 1985	Robert F. Overmyer Frederick D. Gregory Don L. Lind Norman E. Thagard William E. Thornton Lodewijk van den Berg Taylor Wang	7:0:9	Seventeenth STS flight. Spacelab-3 in cargo bay of shuttle.
Space Shuttle Discovery (STS-51G)	Jun. 17, 1985	Daniel C. Brandenstein John O. Creighton Shannon W. Lucid John M. Fabian Steven R. Nagel Patrick Baudry Prince Sultan Salman bin Abdul-Aziz Al-Saud	7:1:39	Eighteenth STS flight. Three communications satellites. One reusable payload, Spartan-l. First U.S. flight with French and Saudi Arabian crew members.
Space Shuttle Challenger (STS-51F)	Jul. 29, 1985	Charles G. Fullerton Roy D. Bridges Karl C. Henize Anthony W. England F. Story Musgrave Loren W. Acton John-David F. Bartoe	7:22:45	Nineteenth STS flight. Spacelab-2 in cargo bay.
Space Shuttle Discovery (STS-51I)	Aug. 27, 1985	Joe H. Engle Richard O. Covey James D. van Hoften William F. Fisher John M. Lounge	7:2:18	Twentieth STS flight. Launched three communications satellites. Repaired Syncom IV-3.
Space Shuttle Atlantis (STS-51J)	Oct. 3, 1985	Karol J. Bobko Ronald J. Grabe	4:1:45	Twenty-first STS flight. Dedicated DOD mission.

		Robert A. Stewart David C. Hilmers William A. Pailes		
Space Shuttle Challenger (STS-61A)	Oct. 30, 1985	Henry W. Hartsfield Steven R. Nagel Bonnie J. Dunbar James F. Buchli Guion S. Bluford, Jr. Ernst Messerschmid (FRG) Reinhard Furrer (FRG) Wubbo J. Ockels (ESA)	7:0:45	Twenty-second STS flight. Dedicated German Spacelab D-1 in shuttle cargo bay.
Space Shuttle Atlantis (STS-61B)	Nov. 27, 1985	Brewster H. Shaw Bryan D. O'Connor Mary L. Cleve Sherwood C. Spring Jerry L. Ross Rudolfo Neri Vela Charles D. Walker	6:22:54	Twenty-third STS flight. Launched three communications satellites. First flight of Mexican astronaut (Neri Vela).
Space Shuttle Columbia (STS-61C)	Jan. 12, 1986	Robert L. Gibson Charles F. Bolden Jr. Franklin Chang-Diaz Steve A. Hawley George D. Nelson Roger Cenker Bill Nelson	6:2:4	Twenty-fourth STS flight. Launched one communications satellite. First member of U.S. House of Representatives in space (Bill Nelson).
Space Shuttle Challenger (STS-51L)	Jan. 28, 1986	Francis R. Scobee Michael J. Smith Judith A. Resnik Ronald E. McNair	0:0:73	Twenty-fifth STS flight. Lost in explosion of liquid hydrogen tank 73 seconds into launch.

		Ellison S. Onizuka Gregory B. Jarvis Christa McAuliffe		
Space Shuttle Discovery (STS-26)	Sep. 29, 1988	Frederick H. Hauck Richard O. Covey John M. Lounge David C. Hilmers George D. Nelson	4:1:0	Twenty-sixth STS flight. Launched TDRS 3.
Space Shuttle Atlantis (STS-27)	Dec. 2, 1988	Robert "Hoot" Gibson Guy S. Gardner Richard M. Mullane Jerry L. Ross William M. Shepherd	4:9:6	Twenty-seventh STS flight. Dedicated DOD mission.
Space Shuttle Discovery (STS-29)	Mar. 13, 1989	Michael L. Coats John E. Blaha James P. Bagian James F. Buchli Robert C. Springer	4:23:39	Twenty-eighth STS flight. Launched TDRS-4.
Space Shuttle Atlantis (STS-30)	May 4, 1989	David M. Walker Ronald J. Grabe Nomman E. Thagard Mary L. Cleave Mark C. Lee	4:0:57	Twenty-ninth STS flight. Venus orbiter Magellan launched.
Space Shuttle Columbia (STS-28)	Aug. 8, 1989	Brewster H. Shaw Richard N. Richards James C. Adamson David C. Leestma Mark N. Brown	5:1:0	Thirtieth STS flight. Dedicated DOD mission.
Space Shuttle Atlantis (STS-34)	Oct. 18, 1989	Donald E. Williams Michael J. McCulley Shannon W. Lucid Franklin R.	4:23:39	Thirty-first STS flight. Launched Jupiter probe and orbiter Galileo.

		Chang-Diaz Ellen S. Baker		
Space Shuttle Discovery (STS-33)	Nov. 23, 1989	Frederick D. Gregory John E. Blaha Kathryn C. Thornton F. Story Musgrave Manley L. "Sonny" Carter	5:0:7	Thirty-second STS flight. Dedicated DOD mission.
Space Shuttle Columbia (STS-32)	Jan. 9, 1990	Daniel C. Brandenstein James D. Wetherbee Bonnie J. Dunbar Marsha S. Ivins G. David Low	10:21:0	Thirty-third STS flight. Launched Syncom IV-5 and retrieved Long Duration Exposure Facility (LDEF).
Space Shuttle Atlantis (STS-36)	Feb. 28, 1990	John O. Creighton John H. Casper David C. Hilmers Richard H. Mullane Pierre J. Thuot	4:10:19	Thirty-fourth STS flight. Dedicated DOD mission.
Space Shuttle Discovery (STS-31)	Apr. 24, 1990	Loren J. Shriver Charles F. Bolden, Jr. Steven A. Hawley Bruce McCandless II Kathryn D. Sullivan	5:1:16	Thirty-fifth STS flight. Launched Hubble Space Telescope (HST).
Space Shuttle Discovery (STS-41)	Oct. 6, 1990	Richard N. Richards Robert D. Cabana Bruce E. Melnick William M. Shepherd Thomas D. Akers	4:2:10	Thirty-sixth STS flight. Launched Ulysses spacecraft to investigate interstellar space and the Sun.
Space Shuttle Atlantis (STS-38)	Nov. 15, 1990	Richard O. Covey Frank L. Culbertson, Jr.	4:21:55	Thirty-seventh STS flight. Dedicated DOD mission.

		Charles "Sam" Gemar Robert C. Springer Carl J. Meade		
Space Shuttle Columbia (STS-35)	Dec. 2, 1990	Vance D. Brand Guy S. Gardner Jeffrey A. Hoffman John M. Lounge Robert A. R. Parker	8:23:5	Thirty-eighth STS flight. ASTRO-1 in cargo bay.
Space Shuttle Atlantis (STS-37)	Apr. 5, 1991	Steven R. Nagel Kenneth D. Cameron Linda Godwin Jerry L. Ross Jay Apt	6:0:32	Thirty-ninth STS flight. Launched Gamma Ray Observatory to measure celestial gamma-rays.
Space Shuttle Discovery (STS-39)	Apr. 28, 1991	Michael L. Coats Blaine Hammond, Jr. Gregory L. Harbaugh Donald R. McMonagle Guion S. Bluford, Jr. Lacy Veach Richard J. Hieb	8:7:22	Fortieth STS flight. Dedicated DOD mission.
Space Shuttle Columbia (STS-40)	Jun. 5, 1991	Bryan D. O'Conner Sidney M. Gutierrez James P. Bagian Tamara E. Jernigan M. Rhea Seddon Francis A. "Drew" Gaffney Millie Hughes-Fulford	9:2:15	Forty-first STS flight. Carried Spacelab Life Sciences (SLS-1) in cargo bay.
Space Shuttle Atlantis (STS-43)	Aug. 2, 1991	John E. Blaha Michael A. Baker Shannon V. Lucid James C. Adamson	8:21:21	Forty-second STS flight. Launched fourth Tracking and Data Relay Satellite (DTRS-5).

Space Shuttle Discovery (STS-48)	Sep. 12, 1991	John Creighton Kenneth Reightler, Jr. Charles D. Gemar James F. Buchli Mark N. Brown	5:8:28	Forty-third STS flight. Launched Upper Atmosphere Research Satellite (UARS).
Space Shuttle Atlantis (STS-44)	Nov. 24, 1991	Frederick D. Gregory Tom Henricks Jim Voss Story Musgrave Mario Runco, Jr. Tom Hennen	6:22:51	Forty-fourth STS flight. Launched Defense Support Program (DSP) satellite.
Space Shuttle Discovery (STS-42)	Jan. 22, 1992	Ronald J. Grabe Stephen S. Oswald Norman E. Thagard David C. Hilmers William F. Readdy Roberta L. Bondar Ulf Merbold (ESA)	8:1:12	Forty-fifth STS flight. Carried International Microgravity Laboratory-1 in cargo bay.
Space Shuttle Atlantis (STS-45)	Mar. 24, 1992	Charles F. Bolden Brian Duffy Kathryn D. Sullivan David C. Leestma Michael Foale Dirk D. Frimout Byron K. Lichtenberg	9:0:10	Forty-sixth STS flight. Carried Atmospheric Laboratory for Applications and Science (ATLAS-1).
Space Shuttle Endeavour (STS-49)	May 7, 1992	Daniel C. Brandenstein Kevin P. Chilton Richard J. Hieb Bruce E. Melnick Pierre J. Thuot Kathryn C. Thornton	8:16:17	Forty-seventh STS flight. Re-boosted a crippled INTELSAT VI communications satellite.

		Thomas D. Akers		
Space Shuttle Columbia (STS-50)	Jun. 25, 1992	Richard N. Richards Kenneth D. Bowersox Bonnie Dunbar Ellen Baker Carl Meade	13:19:30	Forty-eighth STS flight. Carried U.S. Microgravity Laboratory-1.
Space Shuttle Atlantis (STS-46)	Jul. 31, 1992	Loren J. Shriver Andrew M. Allen Claude Nicollier (ESA) Marsha S. Ivins Jeffrey A. Hoffman Franklin R. Chang-Diaz Franco Malerba (Italy)	7:23:16	Forty-ninth STS flight. Deployed Tethered Satellite System-1 and Eureka-1.
Space Shuttle Endeavour (STS-47)	Sep. 12, 1992	Robert L. Gibson Curtis L. Brown, Jr. Mark C. Lee Jerome Apt N. Jan Davis Mae C. Jemison Mamoru Mohri (Japan)	7:22:30	Fiftieth STS flight. Carried Spacelab J. Jemison first African American woman to fly in space. Mohri first Japanese to fly on NASA spacecraft. Lee and Davis first married couple in space together.
Space Shuttle Columbia (STS-52)	Oct. 22, 1992	James D. Wetherbee Michael A. Baker William M. Shepherd Tamara E. Jernigan Charles Lacy Veach	9:56:13	Fifty-first STS flight. Deployed LAGEOS-Satellite, materials experiments, USMP.
Space Shuttle Discovery (STS-53)	Dec. 2, 1992	David M. Walker Robert D. Cabana Guion S. Bluford, Jr. James S. Voss	7:7:19	Fifty-second STS flight. Deployed the last major DOD classified payload planned for Shuttle (DOD1) with ten different secondary payloads.

		Michael Richard Clifford		
Space Shuttle Endeavour (STS-54)	Jan. 13, 1993	John H. Casper Donald R. McMonagle Gregory J. Harbaugh Mario Runco, Jr. Susan J. Helms	6:23:39	Fifty-third STS flight. Deployed Tracking and Data Relay Satellite-6. Operated Diffused X-ray Spectrometer Hitchhiker experiment to collect data on stars and galactic gases.
Space Shuttle Discovery (STS-56)	Apr. 8, 1993	Kenneth D. Cameron Stephen S. Oswald Michael Foale Kenneth D. Cockerell Ellen Ochoa	9:6:9	Fifty-fourth STS flight. Completed second flight of Atmospheric Laboratory for Applications and Science and deployed SPARTAN-201.
Space Shuttle Columbia (STS-55)	Apr. 26, 1993	Steven R. Nagel Terence T. Henricks Jerry L. Ross Charles J. Precourt Bernard A. Harris, Jr. Ulrich Walter (Germany) Hans W. Schlegel (Germany)	9:23:39	Fifty-fifth STS flight. Completed second German microgravity research program in Spacelab D-2.
Space Shuttle Endeavour (STS-57)	Jun. 21, 1993	Ronald J. Grabe Brian J. Duffy David Low Nancy J. Sherlock Peter J. K. Wisoff Janice E. Voss	9:23:46	Fifty-sixth STS flight. Carried SPACEHAB commercial payload module and retrieved European Retrievable Carrier in orbit since Aug. 1992.
Space Shuttle Discovery (STS-51)	Sep. 12, 1993	Frank L. Culbertson, Jr. William F. Readdy James H. Newman Daniel W. Bursch	9:20:11	Fifty-seventh STS flight. Deployed ACTS satellite to serve as testbed for new communications satellite technology and U.S./German ORFEUS/SPAS.

		Carl E. Walz		
Space Shuttle Columbia (STS-58)	Oct. 18, 1993	John E. Blaha Richard A. Searfoss M. Rhea Seddon Shannon W. Lucid David A. Wolf William S. McArthur Martin J. Fettman	14:0:29	Fifty-eighth STS flight. Carried Spacelab Life Sciences mission to determine the effects of microgravity on human and animal subjects.
Space Shuttle Endeavour (STS-61)	Dec. 2, 1993	Richard O. Covey Kenneth D. Bowersox Tom Akers Jeffrey A. Hoffman Kathryn C. Thornton Claude Nicollier F. Story Musgrave	10:19:58	Fifty-ninth STS flight. Restored planned scientific capabilities and reliability of the Hubble Space Telescope.
Space Shuttle Discovery (STS-60)	Feb. 3, 1994	Charles F. Bolden, Jr. Kenneth S. Reightler, Jr. N. Jan Davis Ronald M. Sega Franklin R. Chang-Diaz Sergei K. Krikalev	8:7:9	Sixtieth STS flight. Carried the Wake Shield Facility to generate new semi-conductor films for advanced electronics. Also carried SPACEHAB. Krikalev's presence signified a new era in cooperation in space between Russia and the U.S.
Space Shuttle Columbia (STS-62)	Mar. 9, 1994	John H. Casper Andrew M. Allen Pierre J. Thuot Charles D. Gemar Marsha S. Ivins	13:23:17	Sixty-first STS flight. Carried U.S. Microgravity Payload-2 to conduct experiments in materials processing, biotechnology, and other areas.
Space Shuttle Atlantis (STS-59)	Apr. 9, 1994	Sidney M. Gutierrez Kevin P. Chilton Jerome Apt Michael R. Clifford	11:5:50	Sixty-second STS flight. Carried the Space Radar Laboratory-1 to gather data on the Earth and the effects humans have on its carbon, water, and energy cycles.

		Linda M. Godwin Thomas D. Jones		
Space Shuttle Columbia (STS-65)	Jul. 8, 1994	Robert D. Cabana James D. Halsell, Jr. Richard J. Hieb Carl E. Walz Leroy Chiao Donald A. Thomas Chiaki Naito-Mukai	14:17:55	Sixty-third STS flight. Carried International Microgravity Laboratory-2 to conduct research into the behavior of materials and life in near weightlessness.
Space Shuttle Endeavour (STS-64)	Sep. 9, 1994	Richard N. Richards Blaine Hammond, Jr. J. M. Linenger Susan J. Helms Carl J. Meade Mark C. Lee	10:22:50	Sixty-fourth STS flight. Technology Experiment to perform atmospheric research. Included the first untethered space walk by astronauts in over ten years.
Space Shuttle Discovery (STS-68)	Sep. 30, 1994	Michael A. Baker Terrence W. Wilcutt Thomas D. Jones Steven L. Smith Daniel W. Bursch Peter J. K. Wisoff	11:5:36	Sixty-fifth STS flight. Used Space Radar Laboratory-2 to provide scientists with data to help distinguish human-induced environmental change from other natural forms of change.
Space Shuttle Columbia (STS-66)	Nov. 3, 1994	Donald R. McMonagle Curtis L. Brown, Jr. Ellen Ochoa Joseph R. Tanner Jean-Francois Clervoy (ESA) Scott E. Parazynski	10:22:34	Sixty-sixth STS flight. Three main payloads: third Atmospheric Laboratory for Applications and Science (ATLAS-3), the first Cryogenic Infrared Spectrometers and Telescopes for the Atmosphere-Shuttle Pallet Satellite (CRISTA-SPAS-1), and the Shuttle Solar (SSBUV) spectrometer. Astronauts also conducted experiments.
Space Shuttle Discovery (STS-59)	Feb. 3, 1995	James D. Wetherbee Eileen M. Collins Bernard A. Harris, Jr. Michael Foale	8:6:28	Sixty-seventh STS flight. First close encounter in nearly 20 years with Russian spacecraft, close flyby of Russian Space Station Mir.

		Janice E. Voss Vladmir G. Titov (RSA)		
Space Shuttle Endeavour (STS-67)	Mar. 2, 1995	Stephen S. Oswald William G. Gregory John M. Grunsfeld Wendy B. Lawrence Tamara E. Jernigan Ronald A. Parise Samuel T. Durrance	16:15:8	Sixty-eighth STS flight. Longest flight to date.
Space Shuttle Atlantis (STS-71)	Jun. 27, 1995	Robert L. Gibson Charles J. Precourt Ellen S. Baker Gregory Harbaugh Bonnie J. Dunbar Anatoly Y. Solovyev (RSA) (Left on Mir) Nikolai M. Budarin (RSA) (Left on Mir) Vladimir N. Dezhurov (RSA) (Returned from Mir) Gennady M. Strekalov (RSA) (Returned from Mir) Norman Thagard (Returned from Mir)	9:19:22	Sixty-ninth STS flight. Docked with Russian Space Station Mir and exchanged crews.
Space Shuttle Discovery (STS-70)	Jul. 13, 1995	Terence Hendricks Kevin R. Kregel Nancy J. Currie Donald A. Thomas Mary Ellen Weber	8:22:20	Seventieth STS flight. Deployed TDRS satellite.

Space Shuttle Endeavour (STS-69)	Sep. 7, 1995	David M. Walker Kenneth D. Cockrell James S. Voss James H. Newman Michael L. Gernhardt	10:20:28	Seventy-first STS flight. Deployed Wake Shield Facility and SPARTAN 201-03.
Space Shuttle Columbia (STS-73)	Oct. 20, 1995	Kenneth D. Bowersox Kent V. Rominger Catherine G. Coleman Michael Lopez-Alegria Kathryn C. Thornton Fred W. Leslie Albert Sacco, Jr.	15:21:52	Seventy-second STS flight. Carried out microgravity experiments with the U.S. Microgravity Laboratory (USML-2) payload.
Space Shuttle Atlantis (STS-74)	Nov. 12, 1995	Kenneth D. Cameron James D. Halsell, Jr. Chris A. Hadfield (CSA) Jerry L. Ross William S. McArthur, Jr.	8:4:31	Seventy-third STS flight. Docked with Mir space station as part of International Space Station (ISS) Phase I efforts.
Space Shuttle Endeavour (STS-72)	Jan. 11, 1996	Brian Duffy Brent W. Jett, Jr. Leroy Chiao Winston E. Scott Koichi Wakata (Japan) Daniel T. Barry	8:22:1	Seventy-fourth STS flight. Deployed OAST Flyer. Retrieved previously-launched Japanese Space Flyer Unit satellite. Crew performed spacewalks to build experience for ISS construction.
Space Shuttle Columbia (STS-75)	Feb. 22, 1996	Andrew M. Allen Scott J. Horowitz Jeffrey A. Hoffman Maurizio Cheli (ESA) Claude Nicollier (ESA) Franklin R. Chang-	13:16:14	Seventy-fifth STS flight. Deployed Tethered Satellite System, U.S. Microgravity Payload (USMP-3), and protein crystal growth experiments.

		Diaz Umberto Guidoni (ESA)		
Space Shuttle Atlantis (STS-76)	Mar. 22, 1996	Kevin P. Chilton Richard A. Searfoss Linda M. Godwin Michael R. Clifford Ronald M. Sega Shannon W. Lucid	9:5:16	**Seventy-sixth STS flight. Docked with Mir space station and left astronaut Shannon Lucid aboard Mir. Also carried SPACEHAB module.**
Space Shuttle Endeavour (STS-77)	May 19, 1996	John H. Casper Curtis L. Brown Andrew S. W. Thomas Daniel W. Bursch Mario Runco, Jr. Marc Garneau (CSA)	10:2:30	**Seventy-seventh STS flight. Deployed SPARTAN/Inflatable Antenna Experiment, SPACEHAB, and PAMS-STU payloads.**
Space Shuttle Columbia (STS-78)	Jun. 20, 1996	Terrence T. Henricks Kevin Kregel Richard M. Linnehan Susan J. Helms Charles E. Brady, Jr. Jean-Jacques Favier (CSA) Robert B. Thirsk (ESA)	16:21:48	**Seventy-eighth STS flight. Set Shuttle record for then-longest flight. Carried Life and Microgravity Sciences Spacelab.**
Space Shuttle Atlantis (STS-79)	Sep. 16, 1996	William F. Readdy Terrence W. Wilcutt Jerome Apt Thomas D. Akers Carl E. Walz John E. Blaha Shannon W. Lucid	10:3:19	**Seventy-ninth STS flight. Docked with Mir space station. Picked up astronaut Shannon Lucid and dropped off astronaut John Blaha.**

Space Shuttle Columbia (STS-80)	Nov. 19, 1996	Kenneth D. Cockrell Kent V. Rominger Tamara E. Jernigan Thomas D. Jones F. Story Musgrave	17:3:42	Eightieth STS flight. Longest Mission to date. Deployed ORFEUS-SPAS/Wake Shield.
Space Shuttle Atlantis (STS-81)	Jan. 12, 1997	Michael A. Baker Brent W. Jett, Jr, John M. Grunsfeld Marsha S. Ivins Peter J.K. Wisoff Jerry M. Linenger	10:2:17	Eighty-first STS flight. 5th Mir docking. Night Launch.
Space Shuttle Discovery (STS-82)	Feb. 11, 1997	Kenneth D. Bowersox Scott J. Horowitz Mark C. Lee Steven A. Hawley Gregory J. Harbaugh Specialist Steven L. Smith Joseph R. Tanner	11:2:14	Eighty-second STS mission. Second Hubble Space Telescope servicing mission.
Space Shuttle Columbia (STS-83)	Apr. 4, 1997	James D. Halsell Susan L. Still Janice E. Voss Donald A. Thomas Michael L. Gernhardt Roger Crouch Greg Linteris	5:8:10	Eighty-third STS mission. Microgravity Sciences Laboratory mission. System failed to work properly and mission returned prematurely.
Space Shuttle Atlantis (STS-84)	May 15, 1997	Charles J. Precourt Eileen M. Collins C. Michael Foale Carlos I. Noriega Edward T. Lu	9:6:42	Eighty-fourth STS Mission. 19th Flight OV-104 Atlantis. Sixth Mir docking mission, exchanging crew members.

		Jean-Francois Clervoy (ESA) Elena V. Kondakova (RSA)		
Space Shuttle Columbia (STS-94)	Jul. 1, 1997	James D. Halsell Susan L. Still Janice E. Voss Donald A. Thomas Michael L. Gernhardt Specialist Roger Crouch Greg Linteris	16:12:56	Eighty-fifth STS Mission. 23nd Flight OV-102, Columbia. Microgravity Science Lab-1 Reflight.
Space Shuttle Discovery (STS-85)	Aug. 7, 1997	Curtis L. Brown, Jr. Kent V. Rominger N. Jan Davis Robert L. Curbeam, Jr. Stephen K. Robinson Bjarni Tryggvason (CSA)	11:14:36	Eighty-sixth STS Mission. CRISTA-SPAS-2.
Space Shuttle Atlantis (STS-86)	Sep. 25, 1997	James D. Wetherbee Michael J. Bloomfield Vladimar G. Titov (RSA) Scott E. Parazynski Jean-Loup J.M. Chretien (CNES) Wendy B. Lawrence David A. Wolf	11:14:27	Eighty-seventh STS Mission. Seventh Mir docking mission, exchanging crew members.

Annotated Bibliography

BOOKS

Aeronautics and Space Report of the President. Washington, DC: U.S. Government Printing Office, 1959–1995. Annual series of reports discussing federal government aerospace activities, including key documents and tabular data. Available at Government Repository Libraries.

Armstrong, Neil A., Michael Collins, and Edwin E. Aldrin, Jr. *First on the Moon: A Voyage with Neil Armstrong, Michael Collins and Edwin E. Aldrin, Jr.* Written with Gene Farmer and Dora Jane Hamblin; epilogue by Arthur C. Clarke. Boston: Little, Brown, 1970. The "official" memoir of the *Apollo 11* landing mission to the moon in 1969.

Astronautics and Aeronautics: A Chronology of Science, Technology, and Events. Washington, DC: National Aeronautics and Space Administration, 1962–1990. A multi-volume series of chronologies covering the period 1915–1985.

Atkinson, Joseph D., Jr., and Jay M. Shafritz. *The Real Stuff: A History of the NASA Astronaut Requirements Program.* New York: Praeger, 1985. An overview of the selection of the first ten groups of NASA astronauts through 1984; places heavy emphasis on the criteria for selection and the procedures used in selecting astronauts.

Atwill, William D. *Fire and Power: The American Space Program as Postmodern Narrative.* Athens: University of Georgia Press, 1994. This unique study comments on the development of space exploration as a "big science" program that masked the weaknesses of a "bankrupt" American society.

Benson, Charles D., and William Barnaby Faherty. *Moonport: A History of Apollo Launch Facilities and Operations.* Washington, DC: NASA Special Publication 4204, 1978. A history of the design and construction of the lunar facilities at Kennedy Space Center.

Bergaust, Erik. *Murder on Pad 34.* New York: G. P. Putnam's Sons, 1968. A highly critical account of the investigation of the *Apollo 204* accident in January

1967 that killed astronauts Virgil I. "Gus" Grissom, Roger Chaffee, and Edward White.

———. *Wernher von Braun*. Washington, DC: National Space Institute, 1976. The first full-length biography of the German aerospace engineer and rocket developer who emigrated to the United States in 1945 and is recognized as the most effective advocate of space exploration in the 1950s and 1960s.

Bilstein, Roger E. *Flight in America: From the Wrights to the Astronauts*. Baltimore, MD: Johns Hopkins University Press, 1984; paperback reprint, 1994. A synthesis of the origins and development of aerospace activities in America. This is the book to start with, in any investigation of air and space activities.

———. *Others of Magnitude: A History of the NACA and NASA, 1915–1990*. Washington, DC: NASA Special Publication 4406, 1989. A nonscholarly general history of the National Aeronautics and Space Administration and its predecessor, the National Advisory Committee for Aeronautics.

———. *Stages to Saturn: A Technological History of the Apollo/Saturn Launch Vehicles*. Washington, DC: NASA Special Publication 4206, 1980. This thorough and well-written book gives a detailed but highly readable account of the complex process whereby NASA developed the launch vehicles used in the Apollo program which ultimately sent twelve humans to the moon.

Bonnet, Roger M., and Vittorio Manno. *International Cooperation in Space: The Example of the European Space Agency*. Cambridge, MA: Harvard University Press, 1994. A prize-winning study of the philosophy and inner workings of internationally supported space exploration projects.

Bradbury, Ray, Arthur C. Clarke, Bruce C. Murray, and Carl Sagan. *Mars and the Mind of Man*. New York: Harper and Row, 1973. Analysis by a stellar collection of authors, this book discusses the place of the planet Mars in the mythology and science of humanity from the ancients to the late twentieth century.

Braun, Wernher von, Frederick I. Ordway, III, and Dave Dooling. *History of Rocketry and Space Travel*. New York: Thomas Y. Crowell, 1986 ed. A large-format, illustrated volume that emphasizes the history of U.S. space activities; written by one of the most significant popularizers of space flight, with the assistance of professional writers.

Brooks, Courtney G., James M. Grimwood, and Loyd S., Swenson, Jr. *Chariots for Apollo: A History of Manned Lunar Spacecraft*. Washington, DC: NASA Special Publication 4205, 1979. Based on exhaustive documentary and secondary research as well as 341 interviews, this volume covers the design, development, testing, evaluation, and operational use of the Apollo spacecraft through July 1969.

Bulkeley, Rip. *The Sputniks Crisis and Early United States Space Policy: A Critique of the Historiography of Space*. Bloomington: Indiana University Press, 1991. An important discussion of early efforts to develop civil space policy in the aftermath of the Sputnik crisis of 1957; contains much information relative to the rivalry between the United States and the Soviet Union and how it was affected by the launching of the *Sputnik I* scientific satellite.

Burrows, William E. *Exploring Space: Voyages in the Solar System and Beyond*. New York: Random House, 1990. An insightful discussion of the robotic probes sent to the planets and what scientists learned from these encounters.

Butrica, Andrew J. *To See the Unseen: A History of Planetary Radar Astronomy.* Washington, DC: NASA Special Publication 4218, 1996. An important study that synthesizes the origins and development of a major subelement of space science, ranging from radar imaging of the moon in the 1940s to the *Magellan* radar mapping of Venus in 1989–1990.

Byrnes, Mark E. *Politics and Space: Image Making by NASA.* Westport, CT: Praeger, 1994. A reasoned analysis of the approach taken by the U.S. space agency toward developing its own public image.

CBS News. *10:56:20 PM EDT, 7/20/69: The Historic Conquest of the Moon as Reported to the American People.* New York: Columbia Broadcasting System, 1970. An attempt to capture in print and pictures the reporting on humankind's first landing on the moon during the *Apollo 11* mission.

Chaiken, Andrew. *A Man on the Moon: The Voyages of the Apollo Astronauts.* New York: Viking, 1994. One of the best books on Apollo, this work emphasizes the exploration of the moon by astronauts between 1968 and 1972.

Chaisson, Eric J. *The Hubble Wars: Astrophysics Meets Astropolitics in the Two-Billion-Dollar Struggle Over the Hubble Space Telescope.* New York: HarperCollins, 1994. A provocative but not always reliable discussion of the inner workings and conflicts among scientists, engineers, technological managers, the keepers of space science's image, and public policy advocates concerning the Hubble Space Telescope.

Chang, Iris. *Thread of the Silkworm.* New York: Basic Books, 1996. A biography of H. S. Tsien, a brilliant Chinese rocketeer who studied at Caltech in the 1930s and was deported to the People's Republic of China in 1950 as part of the McCarthy anticommunist purges. Once there, after being jailed for a time, Tsien became the architect of the Chinese ballistic missile program, developing the Silkworm and other rockets for the military.

Clark, Philip S. *The Soviet Manned Space Program.* New York: Crown Publishers, Orion Books, 1988. A general historical work on the Soviet space effort, emphasizing the Cold War rivalries with the United States and how they related to several programs, especially the race to the moon.

Clarke, Arthur C., ed. *The Coming of the Space Age.* New York: Meredith Press, 1967. A compendium of the possibilities of space exploration assembled by the dean of science fiction in the mid-twentieth century.

Collins, Martin J., and Sylvia D. Fries, eds. *A Spacefaring Nation: Perspectives on American Space History and Policy.* Washington, DC: Smithsonian Institution Press, 1991. A collection of presentations from a conference of aerospace exploration that address robotics, human space flight, and technology development.

Collins, Martin J., and Sylvia K. Kraemer, eds. *Space: Discovery and Exploration.* Washington, DC: Hugh Lauter Levin Associates, for the Smithsonian Institution, 1993. A large-format, illustrated history of the U.S. civil space program.

Collins, Michael. *Carrying the Fire: An Astronaut's Journeys.* New York: Farrar, Straus and Giroux, 1974. The first candid book about life as an astronaut, written by the member of the *Apollo 11* crew who remained in orbit around the moon while the other two astronauts visited the lunar surface.

———. *Liftoff: The Story of America's Adventure in Space.* New York: Grove Press, 1988. A general history of the U.S. space program for a popular audience written by a former astronaut.

———. *Mission to Mars: An Astronaut's Vision of Our Future in Space.* New York: Grove Weidenfeld, 1990. An argument on behalf of an aggressive exploration of the red planet, including a review of the earlier advocacies of this effort.

Compton, W. David, and Charles D. Benson. *Living and Working in Space: A History of Skylab.* Washington, DC: NASA Special Publication 4208, 1983. The official NASA history of Skylab, an orbital workshop placed in orbit in the early 1970s.

———. *Where No Man Has Gone Before: A History of Apollo Lunar Exploration Missions.* Washington, DC: NASA Special Publication 4214, 1989. A clearly written account that traces the ways in which scientists transformed a technology program into one that yielded valuable data about the moon.

Cooper, Henry S. F. *Before Lift-Off: The Making of a Space Shuttle Crew.* Baltimore, MD: Johns Hopkins University Press, 1987. Written in a journalistic style, this is an excellent first-person account of training for the 1984 mission of space shuttle STS-41G, which the author participated in.

———. *The Evening Star: Venus Observed.* New York: Farrar, Straus & Giroux, 1993; paperback reprint, Johns Hopkins University Press, 1994. Another of journalist Cooper's excellent first-hand accounts of space exploration, this time observing in the encounter of the *Magellan* spacecraft with Venus in late 1989–1990.

———. *A House in Space.* New York: Holt, Rinehart & Winston, 1976. Another of the journalist's "I was there" accounts, this time about Skylab.

———. *Imaging Saturn: The Voyager Flights to Saturn.* New York: Holt, Rinehart & Winston, 1981. Recounts the dramatic encounter of the *Voyager* spacecraft with Saturn in November 1980 and August 1981, providing a moment-by-moment account of the two flybys and the scientific data they provided.

———. *The Search for Life on Mars: Evolution of an Idea.* New York: Holt, Rinehart & Winston, 1980. A thorough discussion of the lure of Mars for Americans because of the hope that life might exist, or have existed, there.

———. *Thirteen: The Flight That Failed.* New York: Dial Press, 1973. Retells the battle for survival of the *Apollo 13* astronauts after the disabling of the service module when one of its oxygen tanks burst from an electrical malfunction.

Corrigan, Grace. *A Journal for Christa: Christa McAuliffe, Teacher in Space.* Lincoln: University of Nebraska Press, 1993. A touching account of the life of Christa McAuliffe, the teacher who died during the *Challenger* accident of 1986. Written by McAuliffe's mother.

Cortright, Edgar M., ed. *Apollo Expeditions to the Moon.* Washington, DC: NASA Special Publication 350, 1975. A large-formatted volume, with illustrations in both color and black and white, that contains essays by numerous NASA notables.

DeVorkin, David H. *Science with a Vengeance: How the Military Created the U.S.*

Space Sciences after World War II. New York: Springer-Verlag, 1992. Analysis of military experiments with rocketry in the immediate postwar era.

Dick, Steven J. *The Biological Universe: The Twentieth Century Extraterrestrial Life Debate and the Limits of Science*. New York: Cambridge University Press, 1996. A continuation of the author's earlier study, *Plurality of Worlds*, 1982.

Divine, Robert A. *The Sputnik Challenge: Eisenhower's Response to the Soviet Satellite*. New York: Oxford University Press, 1993. Contains insights into the space program as promoted by the Eisenhower White House.

Drake, Frank, and Dava Sobel. *Is Anyone Out There? The Scientific Search for Extraterrestrial Intelligence*. New York: Delacorte Press, 1993. A popular discussion of the issues and efforts surrounding the search for extraterrestrial intelligence.

Emme, Eugene M., ed. *The History of Rocket Technology: Essays on Research, Development, and Utility*. Detroit, MI: Wayne State University Press, 1964. Nearly the only work of detail that surveys rocket technology development in the United States. Although outdated because of events since its publication, the work is unique because it brings together engineers, scientists, and historians to discuss the development of rocketry.

Ezell, Edward Clinton, and Linda Neuman Ezell. *On Mars: Exploration of the Red Planet, 1958–1978*. Washington, DC: NASA Special Publication 4212, 1984. A detailed study of NASA's efforts to send space probes to Mars, culminating with the soft-landing of the two Viking spacecraft in the mid-1970s.

———. *The Partnership: A History of the Apollo-Soyuz Test Project*. Washington, DC: NASA Special Publication 4209, 1978. A detailed study of the effort by the United States and the Soviet Union in the mid-1970s to conduct a joint human space flight.

French, Bevan M., and Stephen P. Maran, eds. *A Meeting with the Universe: Science Discoveries from the Space Program*. Washington, DC: NASA Educational Publication 177, 1981. A discussion of the scientific results of NASA's efforts to explore the planets of the solar system and the wider universe.

Fries, Sylvia D. *NASA Engineers and the Age of Apollo*. Washington, DC: NASA Special Publication 4104, 1992. A sociocultural analysis of a selection of engineers at NASA who worked on Project Apollo; analyzes the manner in which different personalities, perspectives, backgrounds, and priorities influenced the direction of NASA during the 1960s.

Glennan, T. Keith. *The Birth of NASA: The Diary of T. Keith Glennan*. Edited by J. D. Hunley. Washington, DC: NASA Special Publication 4105, 1993. A diary of Eisenhower's NASA administrator; contains a detailed account of the rise and development of the agency between 1958 and the end of 1960.

Goddard, Esther C., ed., and G. Edward Pendray, assoc. ed. *The Papers of Robert H. Goddard*, 3 vols. New York: McGraw-Hill, 1970. A collection of primary source materials that describe the life and work of the premier American rocketry experimenter of the first half of the twentieth century.

Gray, Mike. *Angle of Attack: Harrison Storms and the Race to the Moon*. New York: W. W. Norton, 1992. A lively journalistic account of the career of

Harrison Storms, president of the Aerospace Division of North American Aviation that built the Apollo capsule.

Green, Constance, and Milton Lomask. *Vanguard: A History*. Washington, DC: NASA Special Publication 4202, 1970; reprinted, Smithsonian Institution Press, 1971. An account of the development and operation of what was supposed to be the United States' first orbital satellite in the 1950s.

Hacker, Barton C., and James M. Grimwood. *On Shoulders of Titans: A History of Project Gemini*. Washington, DC: NASA Special Publication 4203, 1977. The official history of the Gemini project conducted by NASA in the mid-1960s.

Hall, R. Cargill. *Lunar Impact: A History of Project Ranger*. Washington, DC: NASA Special Publication 4210, 1977. The official history of the Ranger program to send robotic probes to the moon in the late 1950s and 1960s.

Hallion, Richard P., and Tom D. Crouch, eds. *Apollo: Ten Years since Tranquility Base*. Washington, DC: Smithsonian Institution Press, 1979. A collection of 16 essays developed for the National Air and Space Museum, commemorating the tenth anniversary of the first landing on the moon.

Hanle, Paul A., and Del Chamberlain, eds. *Space Science Comes of Age: Perspectives in the History of the Space Sciences*. Washington, DC: Smithsonian Institution Press, 1981. A collection of essays on all aspects of the space sciences.

Harvey, Brian. *Race into Space: The Soviet Space Programme*. Chichester, England: Ellis Horwood Ltd., 1988. A solid history of the development of the Soviet space program through the mid-1980s. It has several chapters on the race to the moon, describing what information was available before the end of the Cold War in 1989.

Hawthorne, Douglas B. *Men and Women of Space*. San Diego, CA: Univelt, 1992. A useful compendium of biographical information on all the people who have flown in space.

History of Rocketry and Astronautics, 8 vols. San Diego, CA: Univelt, 1986–1995. American Astronautical Society History Series. An eclectic but interesting collection of historical papers on all subjects associated with space exploration delivered at the annual meetings of the International Astronautical Federation.

Hohler, Robert E. *"I Touch the Future . . ."—The Story of Christa McAuliffe*. New York: Random House, 1986. Written by a journalist of the *Concord Monitor*, McAuliffe's hometown newspaper, this is a well-researched and well-written biography of the teacher who died in the *Challenger* accident.

Hudson, Heather E. *Communications Satellites: Their Development and Impact*. New York: Free Press, 1990. No space technology has held more importance for modern America than the communications satellite; the book examines this application and how it has changed life in the twentieth century.

Hufbauer, Karl. *Exploring the Sun: Solar Science since Galileo*. Baltimore, MD: Johns Hopkins University Press, 1991. A prize-winning history of the development of solar science since the fifteenth century, with emphasis on the twentieth-century contribution made possible because of the space age.

Jenkins, Dennis R. *Space Shuttle: The History of Developing the National Space Transportation System*. Osceola, WI: Motorbooks, 1993, 1996. Possibly the best of several popular books that present an overview of the space shuttle and its development and use.

Johnson-Freese, Joan. *Changing Patterns of International Cooperation in Space.* Malabar, FL: Orbit Books, 1990. An exploration of the movement from competition to cooperation in space exploration.

Kauffman, James L. *Selling Outer Space: Kennedy, the Media, and Funding for Project Apollo, 1961–1963.* Tuscaloosa: University of Alabama Press, 1994. A straightforward and helpful history of the public image–building efforts of NASA, and the relation of that image to public policy.

Kay, W. D. *Can Democracies Fly in Space? The Challenge of Revitalizing the U.S. Space Program.* Westport, CT: Praeger, 1995. The answer to the question, of course, is that they can and do; but this analysis suggests that the political requirement to build broad constituencies and to water down proposals to ensure success (as well as to make claims that are beyond reach) makes the effort inefficient, incremental, and always shifting—rather than focused and nonpartisan.

King, Elbert A. *Moon Trip: A Personal Account of the Apollo Program and Its Science.* Houston, TX: University of Houston Press, 1989. A short memoir by a geologist who participated in scientific work on the lunar samples returned by the Apollo missions.

Koppes, Clayton R. *JPL and the American Space Program: A History of the Jet Propulsion Laboratory.* New Haven, CT: Yale University Press, 1982. An institutional history of one of NASA's major centers of space science activities.

Kosloski, Lillian D. *U.S. Space Gear: Outfitting the Astronaut.* Washington, DC: Smithsonian Institution Press, 1993. The only serious history of space suits available.

Krug, Linda T. *Presidential Perspectives on Space Exploration: Guiding Metaphors from Eisenhower to Bush.* New York: Praeger, 1991. An analysis of spaceflight speeches by the presidents, emphasizing their use of romantic analogy and metaphor in their rhetoric about space exploration.

Lambright, W. Henry. *Powering Apollo: James E. Webb of NASA.* Baltimore, MD: Johns Hopkins University Press, 1995. A biography of the NASA administrator between 1961 and 1968, the critical period in which Project Apollo was under way.

Launius, Roger D. *NASA: A History of the U.S. Civil Space Program.* Melbourne, FL: Krieger, 1994. A short book in the Anvil Series, this history of U.S. civilian space efforts is half narrative and half documents.

———, ed. *Organizing for the Use of Space: Historical Perspectives on a Persistent Issue.* San Diego, CA: Univelt, AAS History Series, vol. 18, 1995. A collection of essays related to the logistics of conducting operations in space, organized chronologically and focusing on both civil and military aspects of the U.S. effort.

Lehman, Milton. *This High Man.* New York: Farrar, Straus, 1963. The standard biography of Robert H. Goddard, this book is nonetheless outdated and deserving of replacement.

Levine, Alan J. *The Missile and Space Race.* Westport, CT: Praeger, 1994. Somewhat quirky, this study presents some interesting perspectives on the development of the rivalry between the United States and the Soviet Union in space exploration.

Lewis, Richard S. *The Last Voyage of Challenger*. New York: Columbia University Press, 1988. A follow-on to *Voyages of Columbia*, this book presents (in large-size format with many illustrations) the story of the tragic loss of *Challenger* in 1986.

———. *The Voyages of Apollo: The Exploration of the Moon*. New York: Quadrangle, 1974. An informal yet technical account that covers the background to the Apollo mission seen as an exploration of the moon.

———. *The Voyages of Columbia: The First True Spaceship*. New York: Columbia University Press, 1984. Based on the opinion that the space shuttle is the first true spaceship—one that can be reused and makes access to space more routine—this book describes the development and use of the *Columbia* orbiter.

Ley, Willy. *Rockets, Missiles, and Men in Space*. New York: Viking Press, 1968. The fourth and final edition of 21 printings of the work first published as *Rockets*; emphasizes the possibilities of space flight as reality rather than science fiction. This book became one of the most significant textbooks available in the mid-twentieth century on the possibilities of space travel.

Link, Mae Mills. *Space Medicine in Project Mercury*. Washington, DC: NASA Special Publication 4003, 1965. The first program to launch an American into space was enormously important from the perspective of biomedicine. Could humans survive the rigors of launch on rockets and the harshness of the vacuum of space? This book surveys the questions asked by scientists prior to the project's first launch, and it reviews answers provided by the first astronauts' experiments.

Logsdon, John M. *The Decision to Go to the Moon: Project Apollo and the National Interest*. Cambridge, MA: MIT Press, 1970. A classic study analyzing the political process leading to the U.S. decision to go to the moon in 1961.

———, gen. ed. *Exploring the Unknown: Selected Documents in the History of the U.S. Civil Space Program*, 2 vols. Washington, DC: NASA Special Publication 4407, 1995–1996. An essential reference work containing more than 250 key documents in space policy and its development throughout the twentieth century.

Lovell, Jim, and Jeffrey Kluger. *Lost Moon: The Perilous Voyage of Apollo 13*. Boston: Houghton Mifflin, 1994. After the 1995 feature film *Apollo 13* starring Tom Hanks, no astronaut had more fame than Jim Lovell, commander of the ill-fated mission to the moon in 1970. This book is his own recollection of the mission and the account on which the movie was based.

McConnell, Malcolm. *Challenger: A Major Malfunction*. Garden City, NY: Doubleday, 1987. One of several exposés of NASA's shuttle development and operations management that appeared following the *Challenger* accident.

McCurdy, Howard E. *Inside NASA: High Technology and Organizational Change in the U.S. Space Program*. Baltimore, MD: Johns Hopkins University Press, 1993. Discusses the evolution of NASA's organizational culture, from the creation of the agency to the 1990s, using extensive interviews with key personnel and documentary sources.

———. *The Space Station Decision: Incremental Politics and Technological Choice*.

Baltimore, MD: Johns Hopkins University Press, 1990. A study of the political process that led to the presidential decision in 1984 to develop an orbital space station.

McDougall, Walter A. "*. . . The Heavens and the Earth": A Political History of the Space Age*. New York: Basic Books, 1985. A Pulitzer Prize–winning book that analyzes the race to the moon in the 1960s. The author argues that Apollo prompted the space program to stress engineering over science, competition over cooperation, civilian over military management, and international prestige over practical applications.

Mack, Pamela E. *Viewing the Earth: The Social Construction of Landsat*. Cambridge, MA: MIT Press, 1990. The only substantive study of the origins and development of the first Earth resources monitoring satellite program, launched in the 1970s.

Mailer, Norman. *Of a Fire on the Moon*. Boston: Little, Brown, 1970; London; Weidenfeld & Nicolson, 1970; New York: New American Library, 1971. Written by one of the foremost contemporary American writers, this book examines how the 1960s countercultural mindset met its antithesis, a NASA steeped in middle-class values and reverence for the American flag and culture.

Murray, Bruce C. *Journey into Space: The First Three Decades of Space Exploration*. New York: W. W. Norton, 1989. A discussion of the planetary science program written by the former director of the Jet Propulsion Laboratory.

Murray, Charles A., and Catherine Bly Cox. *Apollo: The Race to the Moon*. New York: Simon and Schuster, 1989. Possibly the best general account of the lunar program, this history uses interviews and documents to reconstruct the stories of the people who participated in Apollo.

NASA Historical Data Book, 4 vols. Washington, DC: NASA Special Publication 4012, 1976–1994. A compendium of basic information on NASA resources, programs, and projects during the period 1958–1978.

Naugle, John E. *First among Equals: The Selection of NASA Space Science Experiments*. Washington, DC: NASA Special Publication 4215, 1991. A discussion of the origin and evolution of NASA's space science experiment decision-making process.

Neal, Valerie, ed. *Where Next, Columbus? The Future of Space Exploration*. New York: Oxford University Press, 1994. A collection of essays linking the voyage of discovery by Columbus with the human exploration of space.

Neal, Valerie, Cathleen S. Lewis, and Frank H. Winter. *Spaceflight: A Smithsonian Guide*. New York: Macmillan, 1995. Provides, with numerous illustrations, a basic history of space exploration by the United States.

Needell, Allan A., ed. *The First 25 Years in Space: A Symposium*. Washington, DC: Smithsonian Institution Press, 1983. A collection of papers from a conference on space exploration held in commemoration of the 1957 launching of *Sputnik I*.

Neufeld, Michael J. *The Rocket and the Reich: Peenemünde and the Coming of the Ballistic Missile Era*. New York: Free Press, 1995. The finest study yet of the German effort in World War II to develop the V-2 ballistic missile.

Newell, Homer E. *Beyond the Atmosphere: Early Years of Space Science*. Washington, DC: NASA Special Publication 4211, 1980. A thoughtful and revealing memoir of space science in NASA during the 1950s and 1960s by the agency's first chief scientist.

Nicks, Orin W., ed. *This Island Earth*. Washington, DC: NASA Special Publication 250, 1970. A classic in the history of government printing, this book contains essays portraying Earth as a fragile lifeboat hanging in the harshness of space. It may have contributed to the rise of the modern environmental movement in the United States, which was also bolstered by the images of Earth first relayed by the space program.

Oberg, James E. *Red Star in Orbit*. New York: Random House, 1981. Written by one of the premier Soviet space watchers, this history of the Soviet space program is among the best published in English prior to the collapse of the Soviet Union in 1991. Based on mostly Western sources, it describes what was then known of the Soviet Union's efforts to land a cosmonaut on the moon before the U.S. Apollo landing in 1969.

Ordway, Frederick I., III, and Mitchell R. Sharpe. *The Rocket Team*. New York: Crowell, 1979. An important, popularly oriented, and somewhat apologetic discussion of the activities of the group of German engineers under the leadership of Wernher von Braun who developed the V-2 during World War II, came to the United States in 1945, and worked at the Marshall Spaceflight Center at Huntsville, Alabama, to develop the Saturn V launch vehicle used in Project Apollo.

Ordway, Frederick I., III, and Randy Lieberman, eds. *Blueprint for Space: From Science Fiction to Science Fact*. Washington, DC: Smithsonian Institution Press, 1992. A collection of essays, accompanied by spectacular artwork and photographs, dealing with the popular culture of space flight in the twentieth century.

Peebles, Curtis. *Watch the Skies! A Chronicle of the Flying Saucer Myth*. Washington, DC: Smithsonian Institution Press, 1994. The best historical study on the debate over the possibility of continuing visitation by extraterrestrials; contains extensive documentation and quotations from official records.

Pisano, Dominick A., and Cathleen S. Lewis, eds. *Air and Space History: An Annotated Bibliography*. New York: Garland Publishing, 1988. The place to start when searching for bibliographical information about any aspect of aerospace history; an exhaustive reference.

Pitts, John A. *The Human Factor: Biomedicine in the Manned Space Program to 1980*. Washington, DC: NASA Special Publication 4213, 1985. Traces the history of space medicine from its early days before the founding of NASA through the decade following the Apollo program.

Reeves, Robert. *The Superpower Space Race: An Explosive Rivalry through the Solar System*. New York: Plenum Press, 1994. Somewhat mistitled (suggesting the broadest possible context of discussion for U.S./USSR rivalry in space exploration), this book is a journalistic account of the robotic race to the various planets by scientists of both nations.

Roman, Peter J. *Eisenhower and the Missile Gap*. Ithaca, NY: Cornell University Press, 1995. A modern analysis, using recently declassified records, of the "missile gap" controversy that arose after the Soviet launch of *Sputnik I*.

Roth, Ladislav E., and Stephen D. Wall. *The Face of Venus: The Magellan Radar-Mapping Mission*. Washington, DC: NASA Special Publication 520, 1995.

A large-format, well-illustrated discussion of the radar-mapping mission conducted by NASA's *Magellan* spacecraft to Venus in 1989–1990.

Sagan, Carl. *Cosmos*. New York: Random House, 1980. Accompanying a PBS science series on the evolution of the universe, this well-illustrated book makes scientific ideas—many coming from research undertaken as part of the space program—both comprehensible and exciting. Although several years old, this is still an outstanding starting point for any research into the universe's origin and evolution.

———. *Pale Blue Dot: A Vision of the Human Future in Space*. New York: Random House, 1994. Probably the most sophisticated articulation of the exploration imperative to appear since Wernher von Braun's work of the 1950s and 1960s.

Sagan, Carl, Frank D. Drake, Ann Druyan, Timothy Ferris, Jon Lomberg, and Linda Salzman Sagan. *Murmurs of Earth: The Voyager Interstellar Record*. New York: Random House, 1978. A discussion of the conceptionalization and carrying out of the effort to place a gold record on the two *Voyager* spacecraft sent outside the solar system in the 1970s. The record contained digital information about planet Earth, including photographs, sounds, music, and greetings in more than 40 languages. It was designed to inform any extraterrestrial intelligence who might encounter it something about this planet and the life that thrives here, and to give that life form a general idea of where Earth is located in space.

Sagan, Carl, and Thornton Page, eds. *UFOs: A Scientific Debate*. New York: W.W. Norton, 1973. A collection of essays about the possibility that there is life beyond Earth in the universe and that intelligent forms of life are visiting Earth. The essays suggest that there is no real proof of the existence of extraterrestrial life, although many people believe it might exist.

Shapland, David, and Michael Rycroft. *Spacelab: Research in Earth Orbit*. Cambridge, England: Cambridge University Press, 1984. A useful discussion of the development and flight of the laboratory built by Europeans for use aboard the space shuttle in Earth-orbit.

Shepard, Alan, and Deke Slayton. *Moonshot: The Inside Story of America's Race to the Moon*. New York: Turner Publishing, 1994. Although it features the recollections of two of the original seven Mercury astronauts chosen in 1959, this book is a disappointing, not very insightful, general history of human space exploration by NASA, from the first flight in 1961 through the last Apollo landing in 1972.

Smith, Robert W. *The Space Telescope: A Study of NASA, Science, Technology, and Politics*. New York: Cambridge University Press, 1989; rev. ed., 1994. A prize-winning history of the development of the Hubble Space Telescope.

Stoker, Carol A., and Carter Emmart, eds. *Strategies for Mars: A Guide to Human Exploration*. San Diego, CA: Univelt, 1996. The most up-to-date and useful of several books related to Mars exploration, this collection of essays provides a rationale, technology assessment, and political analysis of the endeavor through the lens of historical perspective.

Stuhlinger, Ernst, and Frederick I., Ordway, III. *Wernher von Braun: Crusader for Space*, 2 vols. Malabar, FL: Krieger, 1994. A well-illustrated memoir of Wernher von Braun by two co-workers.

Swenson, Loyd S., Jr., James M. Grimwood, and Charles C. Alexander. *This New Ocean: A History of Project Mercury*. Washington, DC: NASA Special Publication 4201, 1966. The official history of Project Mercury, based on extensive research and interviews.

Swift, David W. *SETI Pioneers: Scientists Talk about Their Search for Extraterrestrial Intelligence*. Tucson: University of Arizona Press, 1990. No issue has more affected space exploration than the possibility of making contact with life beyond Earth. This book prints interviews with several scientists and engineers involved in the Search for Extraterrestrial Intelligence, a formal program sponsored by several U.S. organizations.

Tatarewicz, Joseph N. *Space Technology and Planetary Astronomy*. Bloomington: Indiana University Press, 1990. Discusses the interrelationships among space scientists, government patrons, and the politics of "big science" in the 1950s–1980s.

Tomayko, James A. *Computers in Space*. New York: Alpha Books, 1994. A basic history of the capabilities and development of computers used aboard spacecraft for guidance and other computations.

Trento, Joseph J., with reporting and editing by Susan B. Trento. *Prescription for Disaster: From the Glory of Apollo to the Betrayal of the Shuttle*. New York: Crown Publishers, 1987. Not truly an investigation of the *Challenger* accident, this book is an in-depth review of the NASA management and R&D (research and development) system, emphasizing the agency's "fall from grace" after the Apollo program.

Vaughan, Diane. *The Challenger Launch Decision: Risky Technology, Culture, and Deviance at NASA*. Chicago: University of Chicago Press, 1996. The first thorough and scholarly study of events leading to the fateful decision to launch *Challenger* in January 1986; uses sociological and communication theory to piece together the story of America's worst disaster in space flight and to analyze the nature of risk in high-technology enterprises.

Wilford, John Noble. *Mars Beckons: The Mysteries, the Challenges, the Expectations of Our Next Great Adventure in Space*. New York: Alfred A. Knopf, 1990. An explanation of the possibilities of Mars exploration, including a discussion of earlier plans to send humans to the red planet.

Wilhelms, Don E. *To a Rocky Moon: A Geologist's History of Lunar Exploration*. Tucson: University of Arizona Press, 1993. A detailed and contextual account of lunar geology during the 1960s and 1970s, and a less detailed but informative account for the rest of the century.

Winter, Frank H. *Prelude to the Space Age: The Rocket Societies, 1924–1940*. Washington, DC: Smithsonian Institution Press, 1983. A discussion of the private organizations of the early twentieth century devoted to the fostering of space exploration as a fundamental aspect of human destiny.

———. *Rockets into Space*. Cambridge, MA: Harvard University Press, 1990. Possibly the most useful and up-to-date short synthesis of the development of the rocket available in English; written by the rocket curator at the National Air and Space Museum of the Smithsonian Institution.

Wolfe, Tom. *The Right Stuff*. New York: Farrar, Straus & Giroux, 1979. A journalistic account of the first years of space flight, essentially Project Mercury, focusing on the seven Mercury astronauts.

FILM AND VIDEO WORKS

America in Space: The First 40 Years. 1996. Finley-Holiday Film Corp. A 51-minute general video about the history of space exploration by the United States.

Apollo Moon Landings: Out of This World. 1996. Finley-Holiday Film Corp. A 56-minute video providing a general narrative of the Apollo program.

Apollo 13. 1995. Feature film directed by Ron Howard and produced by Brian Graizer, screenplay by William Broyles Jr. and Al Reinert. One of the best feature films ever made about the U.S. space program, this work captures the dynamism and drama of the near-disastrous mission. Tom Hanks as astronaut Jim Lovell and Ed Harris as mission controller Gene Kranz stand out in a fine ensemble cast. Unlike most Hollywood productions, this work paid close attention to historical detail and conveyed the reality of the mission without being overly dramatic or emotional.

Apollo 13—NASA's Historical Film. 1995. Finley-Holiday Film Corp. A 60-minute video history of the mission, originally produced by NASA not long after the flight but re-released in VHS format for educational institutions.

Apollo 13: To the Edge and Back. 1994. WGBH Boston. Written, produced, and directed by Noel Buckner and Rob Whittesey. A 56-minute video history of the mission.

Blue Planet. A 42-minute IMAX film of the National Air and Space Museum. A visually stunning discussion of the ecology of Earth using powerful images from space shuttle missions.

Cosmos. 1977. A 13-part PBS series hosted by Carl Sagan about the relationship of science and civilization and the place of life in the universe. Engaging and thought provoking.

The Dream Is Alive. A 42-minute IMAX film of the National Air and Space Museum. A breathtaking video about space flight, with images from throughout the history of space exploration.

For All Mankind. 1989. An 80-minute documentary film produced and directed by Al Reinert. Deals with the Apollo missions, and uses authentic visuals from the missions and narratives of astronauts on the missions.

Hail Columbia. A 39-minute IMAX film of the National Air and Space Museum. A visually stunning film about the first flight of the space shuttle *Columbia* in 1981.

History of Spaceflight: Reaching for the Stars. 1996. Finley-Holiday Film Corp. A 60-minute video history of NASA, hosted by Alan Shepard.

Hubble Space Telescope: Rescue in Space. 1995. Finley-Holiday Film Corp. A 50-minute video about the December 1994 Hubble servicing mission by the space shuttle.

JPL's First 50 Years. Finley-Holiday Film Corp. A 34-minute video about the origins and development of one of NASA's premier science organizations, the Jet Propulsion Laboratory.

Life on Mars. 1996. Finley-Holiday Film Corp. A 30-minute video about the history of the scientific activities that led to the discovery of possible past life on Mars announced in August 1996.

Mercury & Gemini Spacecraft Missions. Finley-Holiday Film Corp. A 56-minute video about the first piloted American missions into space.

Mission to the Moon. 1986. Signature Productions. Directed by Christine Solinski; written and produced by Blaine Baggett. A 56-minute video about the Apollo program, narrated by Martin Sheen.

Moonshot. 1994. TBS Productions. Produced and directed by Kirk Wolfinger. A 200-minute dramatization, with archival footage, of the history of the human space flight program since the 1950s, hosted by Barry Corbin.

The New Solar System. 1995. Finley-Holiday Film Corp. A 60-minute video about the scientific knowledge gained through space flight since the 1950s, hosted by Joseph Campanella.

One Giant Leap. 1994. Barraclough Carey Productions for Discovery Network. Directed by Steve Riggs, produced by George Carey. A documentary about Project Apollo.

On Robot Wings: Flight thru the Solar System. Finley-Holiday Film Corp. A 35-minute video discussing planetary probes and their discoveries.

The Right Stuff. 1983. Feature film directed by Philip Kaufman and produced by Irwin Winkler and Robert Chartoff, screenplay by Chartoff. A cast of unknown actors at the time depicted the development of aeronautics and astronautics from 1947 through the time of the Mercury program. Scott Glenn, cast as Alan Shepard, played the astronaut perfectly, and Ed Harris as John Glenn captured the essence of being an astronaut. A box-office hit, the film also won an Academy Award for special effects.

Space Shuttle Story: To the Heavens and Beyond. 1996. Finley-Holiday Film Corp. A 60-minute video covering the period 1986–1996; emphasizes the return to flight after the *Challenger* accident and the rendezvous and docking with the Russian space station *Mir* beginning in the mid-1990s.

Space Station MIR: The Best of the Shuttle-Mir Missions. 1996. Finley-Holiday Film Corp. A 50-minute video about the international space station, using spectacular visuals from the docking missions.

To the Moon and Beyond... 1994. SunWest Media Group. A 56-minute video about the Apollo program and the recent history of space exploration.

We Remember: The Space Shuttle Pioneers, 1981–1986. 1996. Finley-Holiday Film Corp. A 60-minute video about the early shuttle program and those who flew on it, emphasizing the seven *Challenger* crew members who died in the January 1986 explosion.

Index

About the Author

ROGER D. LAUNIUS is chief historian of the National Aeronautics and Space Administration in Washington, D.C. He has edited or co-edited several books and articles on aerospace history, including *Spaceflight and the Myth of Presidential Leadership* (1997); *Organizing for the Use of Space: Historical Perspectives on a Persistent Issue* (1995); *NASA: A History of the U.S. Civil Space Program* (1994); *Apollo: A Retrospective Analysis* (1994); and *Apollo 11 at Twenty-Five* (1994). He has also widely written on the history of religion in America.